ENDORSED BY
CAMBRIDGE
International Examinations

OXFORD MECHANICS

1 for Cambridge International AS & A Level

Series editor: David Rayner

**Martin Burgess,
Phil Crossley, Jim Fensom**

Oxford and Cambridge
leading education together

OXFORD

OXFORD
UNIVERSITY PRESS

Great Clarendon Street, Oxford, OX2 6DP, United Kingdom

Oxford University Press is a department of the University of Oxford. It furthers the University's objective of excellence in research, scholarship, and education by publishing worldwide. Oxford is a registered trade mark of Oxford University Press in the UK and in certain other countries

First published in 2015

British Library Cataloguing in Publication Data
Data available

978-0-19-830691-7

10 9 8 7 6 5 4 3 2 1

MIX
Paper from responsible sources
FSC www.fsc.org FSC® C007785

Paper used in the production of this book is a natural, recyclable product made from wood grown in sustainable forests. The manufacturing process conforms to the environmental regulations of the country of origin.

Printed and bound by Bell & Bain Ltd, Glasgow.

The publisher would like to thank Cambridge International Examinations for their kind permission to reproduce past paper questions.

Cambridge International Examinations bears no responsibility for the example answers to questions taken from its past papers which are contained in this publication.

The questions, example answers, marks awarded and comments that appear in this book were written by the authors. In examination, the way marks would be awarded to answers like these may be different.

Acknowledgements
The publishers would like to thank the following for permissions to use their photographs:

Cover: Antun Hirsman/Sutterstock; **p2:** David Wall/Alamy; **p7:** Current Value/Shutterstock; **p10:** Associated Sports Photography/Alamy; **p11t:** Image Source/Alamy; **p11b:** Ian Miles-Flashpoint Pictures/Alamy; **p18:** Maxisport/Shutterstock; **p24:** Hulton Archive/Stringer/Getty; **p30:** andras_csontos/Shutterstock; **p46b:** OUP/Andrew Fosker/Seconds Left/REX; **p46t:** OUP/Suwatchai Pluemruetai; **p47t:** OUP/3D Renderings/Shutterstock; **p47b:** Neil Tingle/Alamy; **p48:** SSPL/Getty Images; **p49:** World History Archive/Alamy; **p50l:** Danny Smythe/Shutterstock; **p50r:** Fedor Selivanov/Shutterstock; **p52t:** Francoise de Valera/Shutterstock; **p52b:** Corepics VOF/Shutterstock; **p57:** Irina Kovancova/Shutterstock; **p59:** Mesut Dogan/Shutterstock; **p62:** Tupungato/Shutterstock; **p72:** Richard Thornton/Shutterstock; **p88t:** Sciepro/Science Photo Library; **p88c:** Sebastien Beaucourt/Look at Sciences/Science Photo Library; **p88b:** Mechanik/Shutterstock; **p89t:** Claus Lunau/Science Photo Library; **p89b:** ZUMA Press, Inc./Alamy; **p90:** Prisma Bildagentur AG/Alamy; **p95:** hektR/Shutterstock; **p98:** Germanskydiver/Shutterstock; **p101:** Mitch Gunn/Shutterstock; **p106:** Natursports/Shutterstock; **p107:** Georgios Kollidas/Shutterstock; **p113:** Jaroslav Pachy sr/Shutterstock; **p118:** Science Museum/Science & Society Picture Library; **p122:** Oleksiy Mark/Shutterstock; **p134t:** RGB Ventures/SuperStock/Alamy; **p134b:** imageBROKER/Alamy; **p135t:** NASA Langley Research Centre; **p135b:** LOC Photo/Alamy;

Contents

1 Straight line motion and graphs 2
1.1 Displacement–time graphs 4
1.2 Velocity–time graphs 8

2 Constant acceleration formulae 18
2.1 Constant acceleration formulae 19
2.2 Vertical motion 24

3 Forces and resultants 30
3.1 Resultants 31
3.2 Components 35
Review exercise A 41
Maths in real-life: Challenging technology in sport 46

4 Newton's laws 48
4.1 Newton's laws 49
4.2 Resolving when on an inclined plane 53
4.3 Multiple forces 55
4.4 Connected particles 57

5 Equilibrium 62
5.1 Three forces acting at a point 63

6 Friction 72
6.1 Rough horizontal surfaces 73
6.2 Rough inclined slope 76
Review exercise B 84
Maths in real-life: Celestial mathematics 88

7 Work and energy 90
7.1 Work 91
7.2 Kinetic energy 94
7.3 Gravitational potential energy 95
7.4 Conservation of energy 96
7.5 The work–energy principle 99

8 Power 106
8.1 Power as rate of doing work 107
8.2 Acceleration and variable resistance 110

9 Variable forces 118
9.1 Using differentiation to describe straight line motion 119
9.2 Using integration to describe straight line motion 123
9.3 The constant acceleration formulae 126
Review exercise C 129
Maths in real-life: Aerodynamics 134

Exam-style paper A 136
Exam-style paper B 138
Answers 140
Index 147

Introduction

About this book

This book has been written to cover the **Cambridge AS and A Level International Mathematics (9709)** course, and is fully aligned to the syllabus.

In addition to the main curriculum content, you will find:

- 'Maths in real-life', showing how principles learned in this course are used in the real world.
- Chapter openers, which outline how each topic in the Cambridge 9709 syllabus is used in real-life.

The book contains the following features:

Notes

Did you know?

Advice on calculator use

Examination advice

Throughout the book, you will encounter worked examples and a host of rigorous exercises. The examples show you the important techniques required to tackle questions. The exercises are carefully graded, starting from a basic level and going up to exam standard, allowing you plenty of opportunities to practise your skills. Together, the examples and exercises put maths in a real-world context, with a truly international focus.

At the start of each chapter, you will see a list of objectives that are covered in the chapter. These objectives are drawn from the Cambridge AS and A Level syllabus. Each chapter begins with a *Before you start* section and finishes with a *Summary exercise* and *Chapter summary*, ensuring that you fully understand each topic.

A *Review exercise* is placed after chapters 3, 6 and 9. There comprise a host of exam questions which cover the topics from the previous 3 chapters and are in no particular order of difficulty.

The answers given at the back of the book are concise. However, when answering exam-style questions, you should show as many steps in your working as possible. All exam-style questions, as well as *Paper A* and *Paper B*, have been written by the authors.

About the authors

Jim Fensom has many years of experience teaching and examining mathematics. He has authored a number of books and is currently mathematics coordinator at Nexus International School in Singapore.

Phil Crossley is currently a senior examiner as well as a teacher at Carre's Grammar School in England. He has many years of experience in teaching and examining mathematics.

Dr Martin Burgess has over nine years' experience in teaching mathematics at secondary level and has also been an expert examiner for an A-level examination board. His Ph.D. is in the field of data mining, specialising in statistical techniques, and he currently works at Nexus International School in Singapore.

Special thanks to James Nicholson for 'Maths in real-life'.

A note from the authors

The aim of this book is to help students prepare for the Mechanics 1 unit of the Cambridge International AS and A Level Mathematics syllabus, though it may also be found to be useful in providing support material for other AS and A Level courses. The book contains a large number of practice questions, many of which are exam-style in addition to questions from past Cambridge examination papers.

In writing the book we have drawn on our experiences of teaching Mathematics over many years, as well as our experience as examiners.

1 Straight line motion and graphs

The longest straight stretch of train track in the world is in Australia. It runs from Ooldea, in South Australia to Loongana, in Western Australia, a distance of 478 km. This section of track is part of the Trans-Australian Railway on which the Indian Pacific line from Sydney, in the East of Australia, to Perth, in the West, runs. It runs though the Nullarbor Plain, an area of flat, almost treeless, arid or semi-arid country that occupies an area of 200 000 square kilometres, The length of the journey is 4352 km one-way, and takes 65 hours. The average speed of train is 85 km/h and its maximum speed is 115 km/h.

Objectives

Sketch and interpret displacement–time and velocity-time graphs, and in particular appreciate that

- the area under a velocity-time graph represents displacement
- the gradient of a displacement–time graph represents velocity
- the gradient of a velocity-time graph represents acceleration.

Before you start

You should know how to:

1. Calculate the area of rectangles, triangles and trapeziums.

e.g.

Area = 2.4 × 5
= 12 cm^2

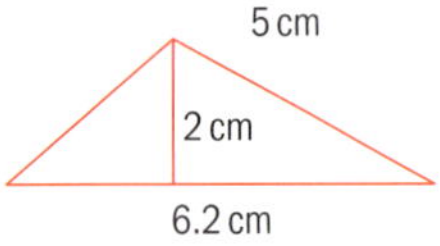

Area = 6.2 × 2
= 12.4 cm^2

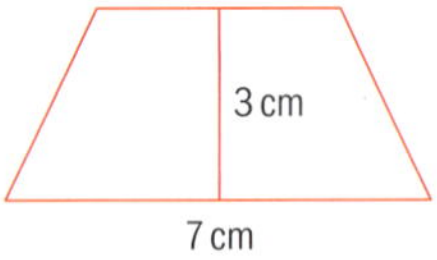

Area = $\frac{1}{2}(4 + 7) \times 3$
= 16.5 cm^2

2. Calculate the gradient of a straight line.

Use gradient = $\frac{y_2 - y_1}{x_2 - x_1}$. e.g. Find the gradient of the line joining (2, 4) and (7, −1)

gradient = $\frac{-1-4}{7-2} = \frac{-5}{5} = -1$

Skills check:

1. Calculate the area of the shape created by the red line of this graph.

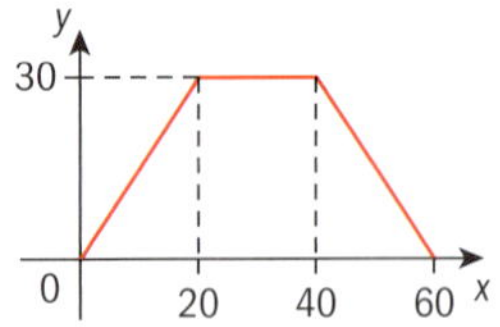

2. Calculate the gradients of these lines.

a)

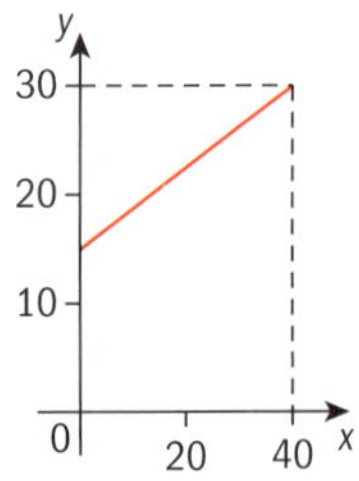

b)

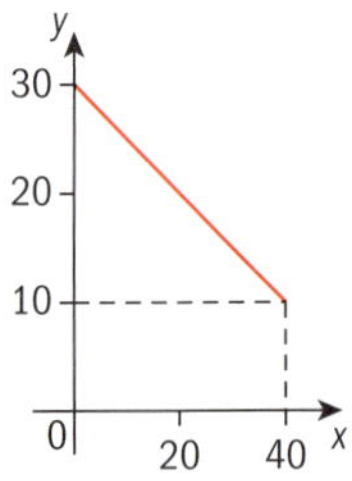

3. Calculate with displacement, velocity, acceleration and time using appropriate units (m, ms^{-1}, ms^{-2}, s).

When velocity is constant then the formulae that connect these quantities are as follows:

displacement = velocity × time e.g. Find the displacement of a particle travelling at 3 ms^{-1} for 5 s.

displacement = 3 × 5 = 15 m

$\text{velocity} = \frac{\text{displacement}}{\text{time}}$ e.g. Find the velocity when a particle has a displacement of 2.4 m in 3 s.

$\text{velocity} = \frac{2.4}{3} = 0.8\ ms^{-1}$

$\text{time} = \frac{\text{displacement}}{\text{velocity}}$ e.g. Find the time taken for a particle to be displaced 15 m with a velocity of 0.6 ms^{-1}.

$\text{time} = \frac{15}{0.6} = 25s$

When acceleration is constant then the formulae that connect these quantities are as follows:

velocity = acceleration × time e.g. Find the change in velocity when a particle accelerates for 2 s at 24 ms^{-2}.

change in velocity = 24 × 2 = 48 ms^{-1}.

$\text{acceleration} = \frac{\text{velocity}}{\text{time}}$ e.g. Find the acceleration when the velocity of a particle changes from 2 ms^{-1} to 10 ms^{-1} in 12 s.

$\text{acceleration} = \frac{10-2}{12} = \frac{2}{3}\ ms^{-2}$

$\text{time} = \frac{\text{velocity}}{\text{acceleration}}$ e.g. find the time taken for a particle to accelerate to a velocity of 8 ms^{-1} from 3 ms^{-1} when its acceleration is 0.1 ms^{-2}.

$\text{time} = \frac{5}{0.1} = 50s$

3. Find the time taken for a particle travelling

a) 30 m at a velocity of 5 ms^{-1}

b) 8 m at a velocity of 0.2 ms^{-1}

c) 5 m at a velocity of 25 ms^{-1}.

4. Find the displacement of a particle travelling with

a) a velocity of 12 ms^{-1} for 12 s

b) a velocity of 0.4 ms^{-1} for 10 s

c) a velocity of 30 ms^{-1} for 0.5 s.

5. Find the velocity of a particle that has

a) a displacement of 24 m in 8 s

b) a displacement of 45 m in 30 s

c) a displacement of 10 m in 50 s.

6. Find the change in velocity when a particle accelerates at

a) 10 ms^{-2} for 10 s

b) 0.2 ms^{-2} for 30 s.

7. Find the acceleration when a particle's velocity changes

a) from 20 ms^{-1} to 50 ms^{-1} in 10 s

b) from 44 ms^{-1} to 32 ms^{-1} in 6 s.

8. Find the time taken for a particle to accelerate from 15 ms^{-1} to 60 ms^{-1} at 15 ms^{-2}.

1.1 Displacement–time graphs

A displacement–time graph is used to show the motion of a **particle**, in one dimension, along a straight line. We first look at examples where motion follows one or more stages of constant velocity, with the particle moving forwards and backwards along the straight line. In displacement–time graphs, time (t) is shown on the horizontal axis. Displacement is often denoted by s.

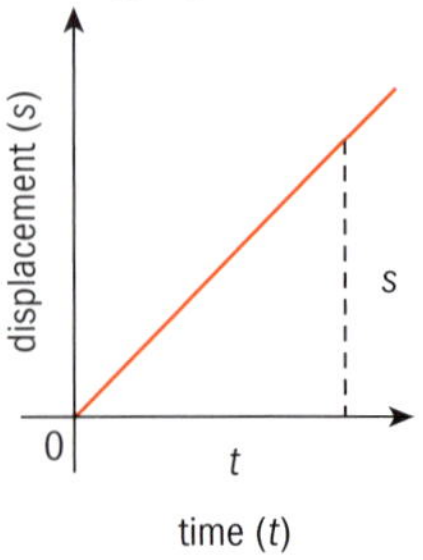

We know that velocity $= \dfrac{\text{displacement}}{\text{time}}$, which can be abbreviated as velocity $= \dfrac{s}{t}$.

Velocity = gradient of the displacement–time graph.

Note that the concept of **kinematics**, or straight-line motion, refers to the motion of a particle. A particle has dimensions so small compared with other lengths that its position in space can be represented by a single point. A **body** is an object made up of particles. However, in Example 1, a body (in this case a car) is treated as a particle for the purpose of the question.

Example 1

A car moves forward on a straight road from a point O, at constant velocity for 20 s, travelling a distance of 60 m. During the next 20 s the car is stationary, remaining 60 m away from O. The car then returns to O, which takes 10 s.

a) Sketch a displacement–time graph of the first 50 s of the car's journey.

b) Use the displacement–time graph to find the velocity of the car during each stage of the journey.

a)

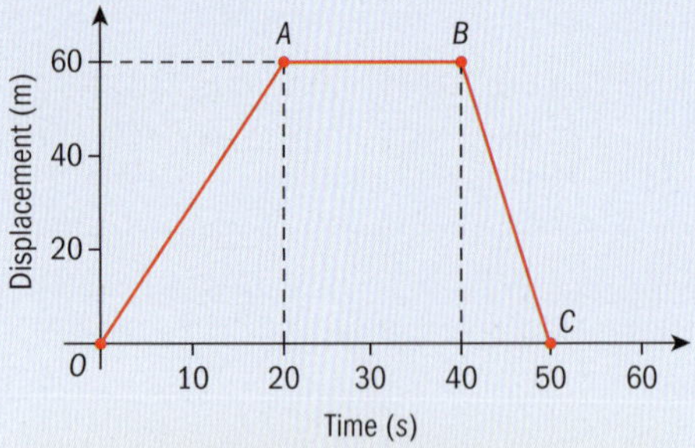

Note: In this example, displacement away from O is regarded as positive; hence, on the return part of the journey both the displacement and the velocity are negative.

b) Since velocity $= \dfrac{\text{displacement}}{\text{time}}$, the gradient of a displacement–time graph is the velocity.

The gradient of OA is $\dfrac{60}{20} = 3$

The velocity in the first 20 s is $3\,\text{ms}^{-1}$.

The gradient of AB is 0.

The velocity in the second 20 s is $0\,\text{ms}^{-1}$.

The gradient of BC is $\dfrac{-60}{10} = -6$

The velocity in the final 10 s is $-6\,\text{ms}^{-1}$.

Exercise 1.1

1. A particle starts from a point O at a velocity of $2\,\text{ms}^{-1}$ for 10 s, rests for 20 s and then returns to O in 5 s.

   a) Sketch the displacement–time graph of the motion of the particle.

   b) What is the velocity of the particle on the return?

2. A car travels along a straight road from a town O. It travels 200 m at a constant velocity of $20\,\text{ms}^{-1}$. It then stops for 5 seconds before returning to the starting point in 8 s.

   a) Sketch a displacement–time graph for the motion of the car.

   b) Calculate the velocity on the return section of the journey.

3. A food container in a sushi restaurant travels along a straight track at a velocity of $0.5\,\text{ms}^{-1}$ for 10 s. It stops for 10 s and then continues on its journey at a velocity of $0.6\,\text{ms}^{-1}$, coming to a halt after 10 s.

   a) Sketch the displacement–time graph for the food container.

   b) Calculate the total distance travelled by the food container.

In the graphs that follow, displacement (s) is given in metres, and time (t) is in seconds.

4. Describe the motion of the particle in the graph.
   What is the velocity of the particle

   a) in the first 10 seconds

   b) between $t = 10$ and $t = 40$

   c) in the last 30 seconds?

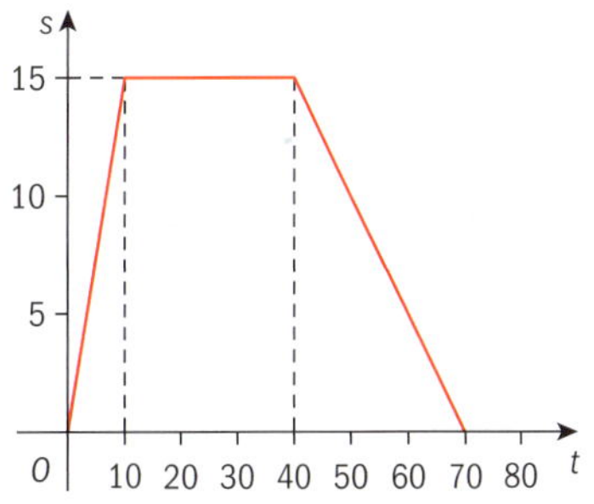

5. Describe the motion of a train shunting along a track, as shown in the accompanying graph.
   What is the velocity of the train

   a) in the first 5 seconds

   b) between $t = 5$ and $t = 20$

   c) in the last 15 seconds?

"Shunt" means to push or pull a train from one line of rails to another.

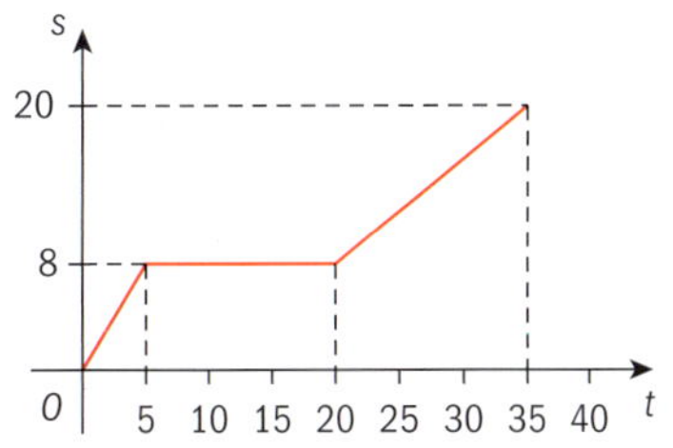

6. Describe the motion of an elevator moving in an elevator shaft, as shown in the graph.

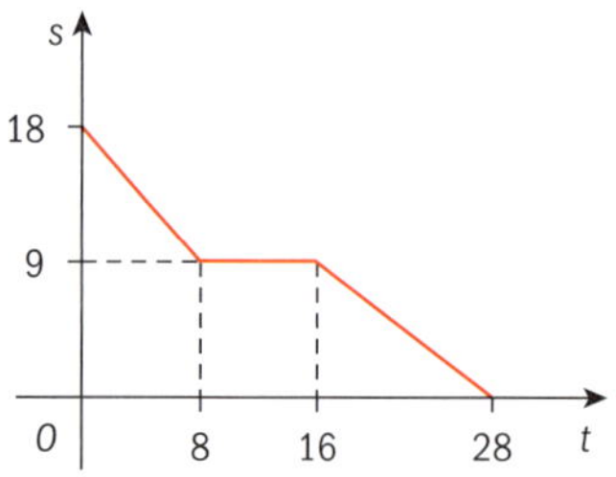

What is the velocity of the elevator

a) in the first 8 seconds

b) between $t = 8$ and $t = 16$

c) in the last 12 seconds?

7. In a factory, a piece of steel travels a distance S from O in 40 s on a conveyor belt and then returns to O 20 s later.

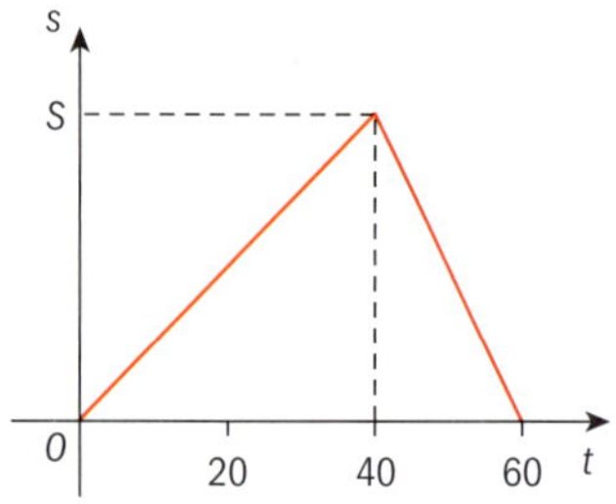

If the initial velocity of the piece of steel is $8\,\text{ms}^{-1}$, calculate

a) the value of S

b) the velocity of the piece of steel between $t = 40$ and $t = 60$.

8. A particle travels 20 m from O in T s. It remains stationary for a further T s, and then returns to O in T s.

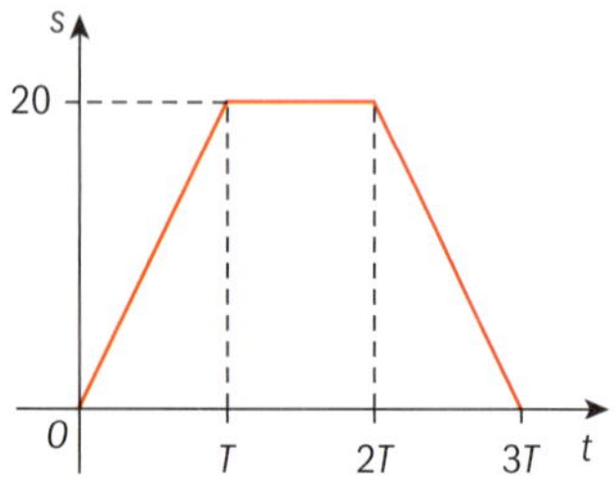

If the initial velocity of the particle is $30\,\text{ms}^{-1}$, calculate

a) the value of T

b) the time taken to complete the whole journey.

9. The displacement of a particle from O is S m in a time $4T$ s. The particle then returns to O as shown in the graph.

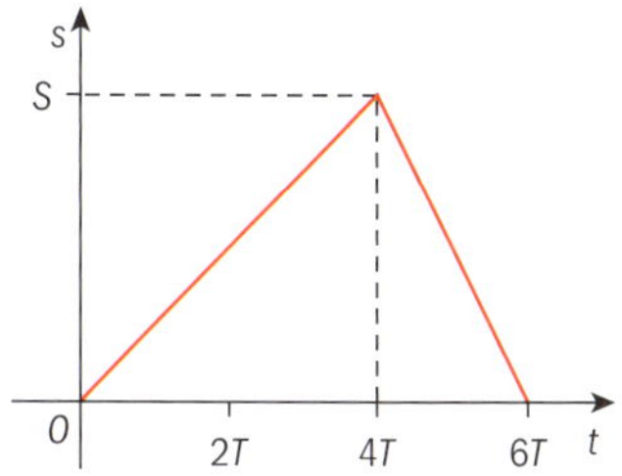

a) If the initial velocity is $4\,\text{ms}^{-1}$, find the velocity on the return.

b) If the total time taken is 27 s, find the value of S.

10. The graph shows the displacement s of a model train moving along a track in time t.

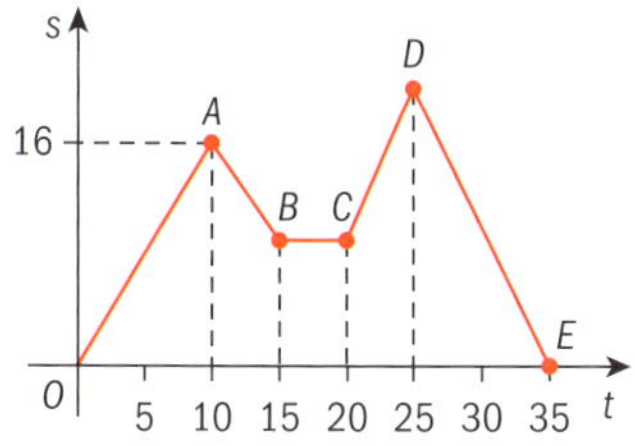

The velocity of the train from O to A is $V\,\text{ms}^{-1}$ and from A to B the velocity is $-V\,\text{ms}^{-1}$. The train is at rest between B and C and between C and D, the velocity is $1.5V\,\text{ms}^{-1}$. Calculate

a) the value of V

b) the displacement from A to B

c) the displacement from C to D

d) the displacement from D to E

e) the velocity between D and E.

Did you know?

The examples in section 1.1 are a simplification of what happens in real-life. In practice, although a particle (or body) can travel at a constant velocity, change in velocity is never instantaneous. It involves **acceleration** or **deceleration**.

Imagine sitting in a car where the velocity changed abruptly. What would happen to your body if it sped up, slowed down, stopped, or changed direction in no time at all?

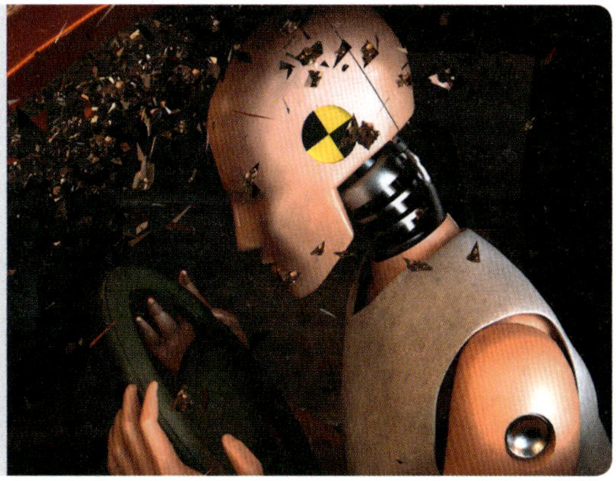

1.2 Velocity–time graphs

A velocity–time graph is also used to show the motion of a particle in one dimension, along a straight line. In the next set of examples, motion follows one or more stages of constant velocity or constant acceleration, with the particle moving forwards and backwards along the straight line. In velocity–time graphs, time (t) is shown on the horizontal axis. Velocity is denoted by v.

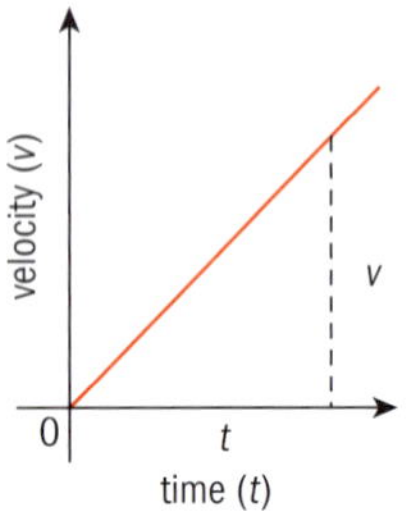

We know that acceleration $= \frac{\text{velocity}}{\text{time}}$, which can be abbreviated as acceleration $= \frac{v}{t}$. Therefore,

> Acceleration is the gradient of the velocity–time graph.

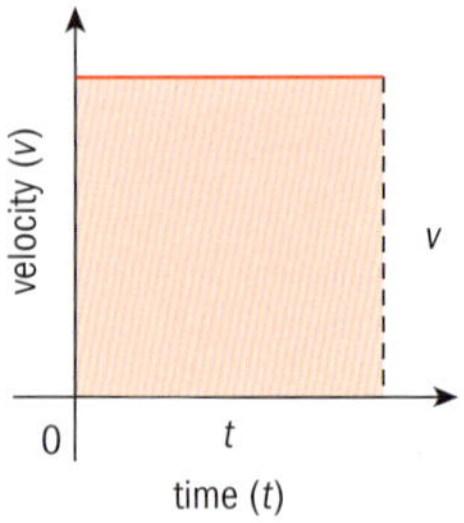

When velocity is constant, displacement = velocity × time, or more simply, $v \times t$. Therefore,

> For constant velocity, displacement is found by calculating the area of the rectangle on a velocity–time graph.

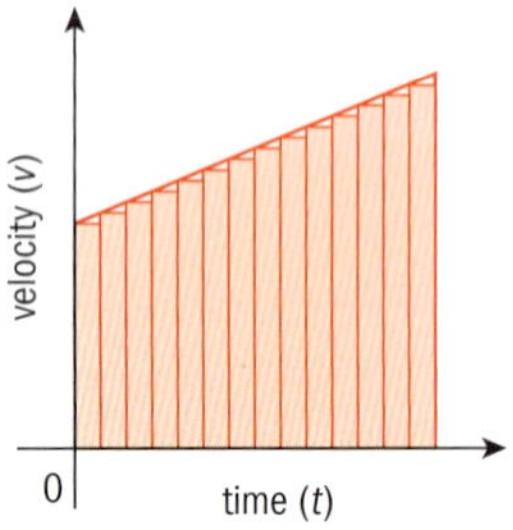

When acceleration is constant, the velocity graph will be a straight line. Consider the area under the graph to be made of a series of very narrow rectangles. The area of each of these rectangles is the displacement of the particle over a very short time. Combining these areas, we get an approximation to the area of the trapezium under the graph, which improves as the time period for each rectangle become less.

> Displacement is the area under the velocity–time graph

Example 2

During the first 10 s of a journey along a straight road, a car accelerates from rest to a velocity of 20 ms^{-1}. It then continues for a further 20 s at constant velocity.

a) Draw the velocity–time graph of the journey.

b) Calculate the acceleration during the first 10 s of the journey.

c) Describe the motion between the first 10s and 30 s of the journey.

d) Calculate the total distance travelled in the first 30 s of the journey.

a)

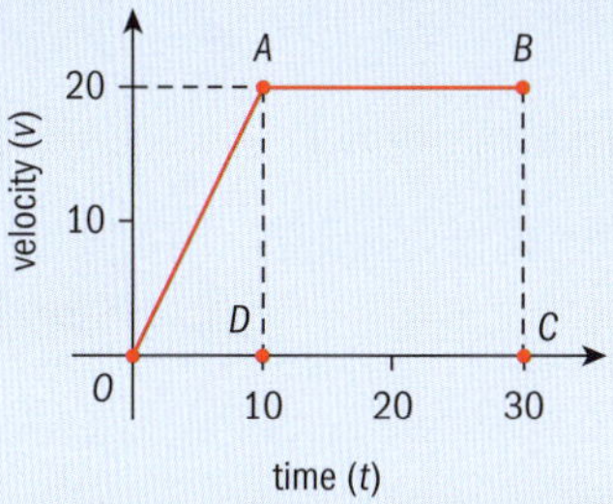

b) Since acceleration $= \frac{\text{velocity}}{\text{time}}$, the gradient of a velocity–time graph is the acceleration.

The gradient of OA is $\frac{20}{10} = 2$.

The acceleration in the first 10 s is 2 ms^{-2}.

c) The gradient of the graph is 0 between A and B. Therefore the acceleration is 0 ms^{-2}. This is constant velocity.

d) Since displacement = velocity × time, the displacement is found by calculating the area under the graph between $t = 0$ and $t = 30$.

Total area = area of OAD + area of $ABCD$

$= \frac{1}{2} \times 10 \times 20 + 20 \times 20$

$= 500$

or

Total area = area of trapezium $OABC$

$= \frac{1}{2}(30 + 20) \times 20$

$= 500$

The displacement in the first 30 s is 500 m.

Note: Velocity away from O is regarded as positive. Lines with a positive gradient show positive acceleration and those with a negative slope show negative acceleration (deceleration).

Hint: Using the trapezium formula is often quicker and more straightforward than breaking the area into simpler shapes.

In a velocity–time graph, velocity can be positive (above the axis) or negative (below the axis). Performing calculations from values in the graph can result in areas that are negative as well, indicating a negative displacement. Care needs to be taken when finding displacement if both positive and negative velocities are involved.

Consider the case of a cricket ball thrown up in the air. There is a positive displacement as the ball travels up, and a negative displacement as it travels back down. Since the ball returns to the point where it starts, the overall displacement is zero. The distance travelled however, is not equal to zero. The difference between displacement (a **vector** quantity) and distance (a **scalar** quantity) will be discussed in Chapter 2.

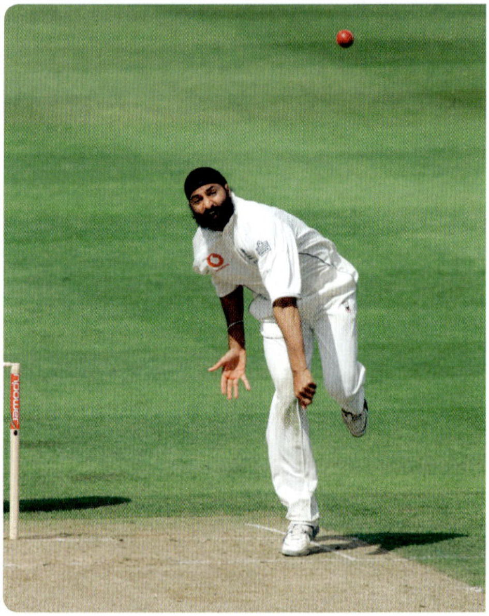

Example 3

A ball is projected up a smooth plane at a velocity of 15 ms^{-1} from a point O. The ball decelerates at constant rate for 6 s. The ball is instantaneously at rest at $t = 3$ s.

a) Draw the velocity–time graph for the ball's journey.

b) Calculate the value of the ball's acceleration.

c) What is the maximum displacement of the ball?

d) What is the total distance travelled by the ball?

a)

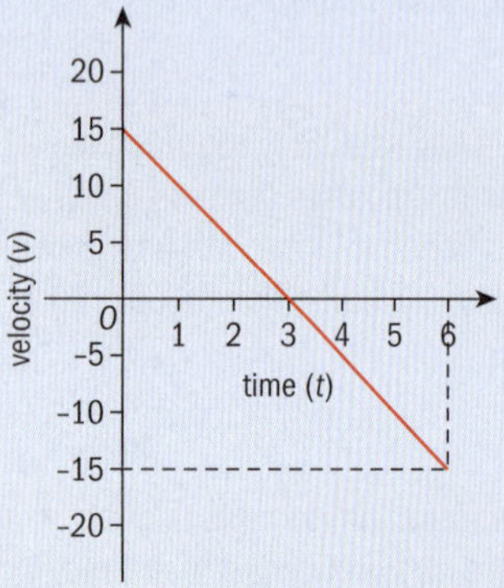

b) Gradient $= \frac{-30}{6} = \frac{-15}{3} = -5$

The acceleration is -5 ms^{-2}.

c) The maximum displacement of the ball occurs when it is at rest, when $t = 3$ s.

Area $= \frac{1}{2} \times 3 \times 15 = 22.5$

The maximum displacement is 22.5 m.

d) Between $t = 3$ and $t = 6$, the displacement is -22.5 m and hence the total displacement of the ball is $22.5 + (-22.5) = 0$ m. This simply means that the ball has returned to its starting position.

The total *distance* travelled is $2 \times 22.5 = 45$ m.

Note that another difference between vector and scalar quantities is between velocity and speed. Consider two cars which hit each other on a highway travelling at the same velocity, and two other cars which hit each other head-on at the same speed.

- **Result**: the first two cars have only minor damage, while the second two have significant damage.

What is a possible explanation for this difference in damage?

- **The reason**: two cars travelling at the same velocity are travelling in the same direction and so they hit each other in a side-on collision, causing less damage. In straight-line motion, speed could be in opposite directions, and a head-on collision would cause much more damage.

Example 4

A car accelerates smoothly from rest for 30 s to a velocity of 20 ms^{-1}. It continues at a steady velocity for 20 s before decelerating to rest in 20 s.

a) Draw the velocity–time graph of the motion of the car in the first 70 s of motion.

b) Calculate the acceleration in the first 30 s and the final 20 s.

c) Calculate the total displacement.

d) Calculate the average speed during the journey.

a)

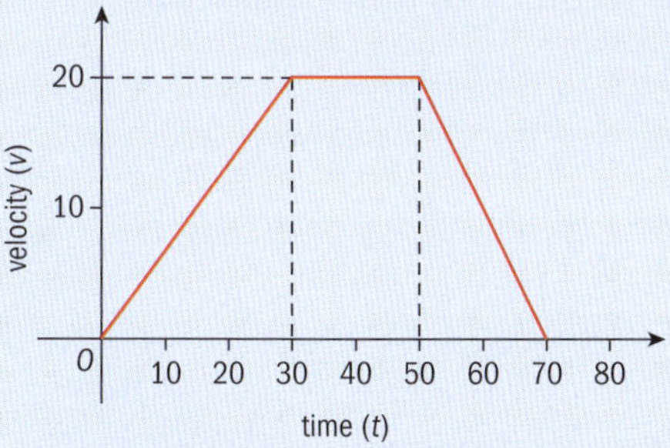

b) Since acceleration $= \frac{\text{velocity}}{\text{time}}$, the gradient of a velocity–time graph is the acceleration.

The gradient in the first 30 s is $\frac{20}{30} = \frac{2}{3} \approx 0.667$

The acceleration in the first 30 s is 0.667 ms^{-2}.

The gradient in the last 20 s is $\frac{-20}{20} = -1$

The acceleration in the last 20 s is -1 ms^{-2}.

▶ Continued on the next page

c) Since displacement = velocity × time, the displacement is found by calculating the area under the graph between $t = 0$ and $t = 70$.

Total area = area of trapezium

$$= \frac{1}{2}(70 + 20) \times 20$$
$$= 900$$

The total displacement is 900 m.

d) $\text{Average speed} = \dfrac{\text{total displacement}}{\text{time taken}}$

$\text{Average speed} = \dfrac{900}{70} \approx 12.9\,\text{ms}^{-1}$

It is a common error not to use the correct formula when calculating average speed. Make sure that you learn this formula:

$\text{average speed} = \dfrac{\text{total displacement}}{\text{time taken}}$

Note: Average speed is *not* found by taking the average of two speeds.

Exercise 1.2

1. A car is travelling at $30\,\text{ms}^{-1}$. It continues at constant velocity for 20 s, and then slows to a halt after a further 10 s. Draw a velocity–time graph to show the motion of the car.

2. A baseball is thrown vertically at an initial velocity of $20\,\text{ms}^{-1}$. The ball decelerates under gravity at a rate of $10\,\text{ms}^{-2}$. Draw a velocity–time graph to show the motion of the baseball from the time it is thrown until it reaches the ground again.

3. A train leaves a station and accelerates for 10 s until it reaches a velocity of $24\,\text{ms}^{-1}$. It then travels at constant velocity for 60 s until it approaches the next station when it decelerates at a rate of $2\,\text{ms}^{-2}$. Draw a velocity–time graph to show the motion of the train.

In the graphs that follow, velocity (v) is given in metres per second, and time (t) in seconds.

4.

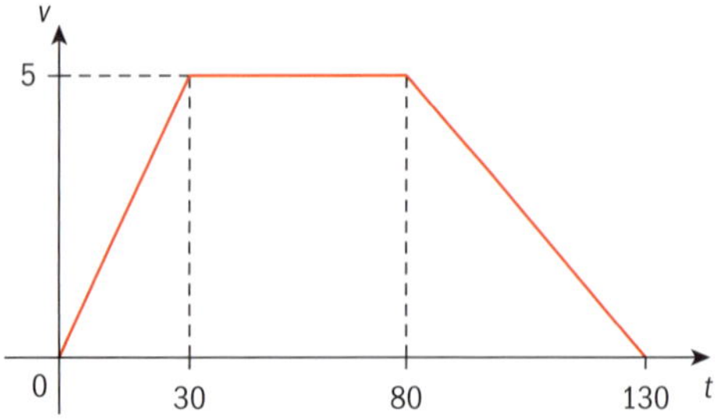

The velocity–time graph shows the motion of boat moving along a straight canal. The boat moves forward for 30 s, accelerating from rest to a velocity of $5\,\text{ms}^{-1}$, then moves for another 50 s at constant velocity and finally decelerates for another 50 s until it is at rest again. Calculate

a) the acceleration when $t = 10$ and when $t = 100$

b) the total distance covered

c) the average speed for the whole journey.

5. The velocity–time graph shows the first 35 s of the motion of a car as it moves onto a highway. In the first 10 s it accelerates from rest to 16 ms^{-1} on the slip road. It then travels for 10 s on the slip road at constant velocity before pulling out onto the highway and accelerating for another 15 s to reach a velocity of 40 ms^{-1}. Find

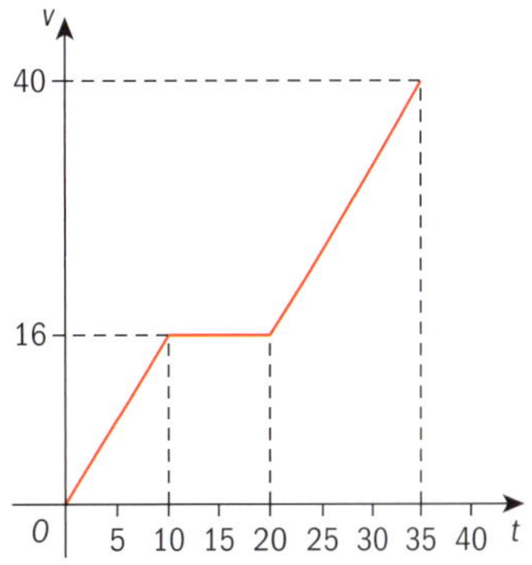

a) the acceleration when $t = 5$ and $t = 30$

b) the distance travelled by the car on the slip road

c) the total distance travelled by the car during the first 35 s.

6.

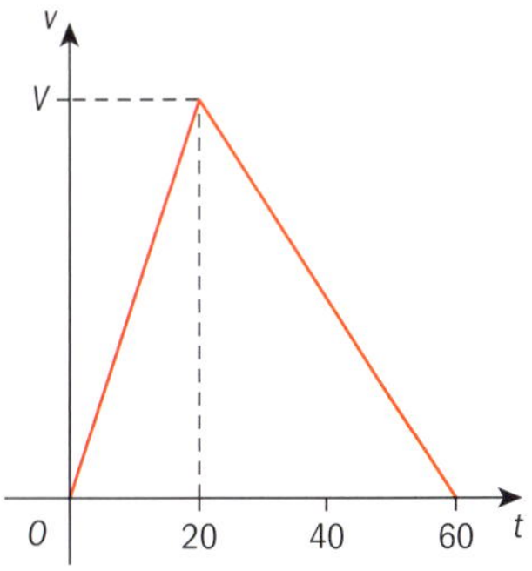

A particle accelerates to a velocity V in 20 s and then decelerates back to rest in 40 s. Find

a) V if the total distance travelled is 450 m

b) the acceleration in the first 20 s

c) the total distance travelled in the first 40 s.

7.

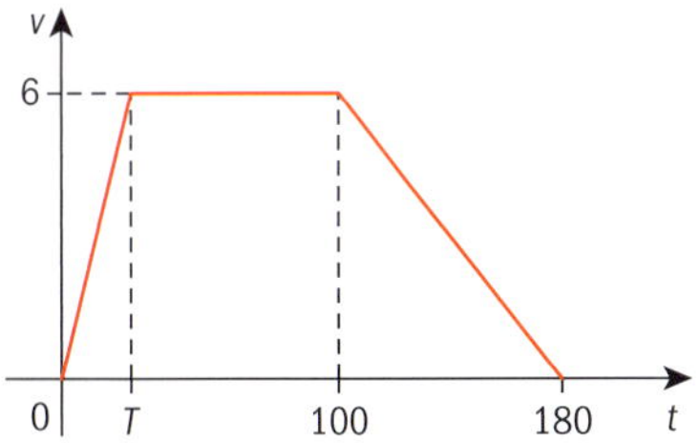

In a mineshaft, an elevator is bringing coal from a coal mine below ground. The motion of the elevator is modelled by three straight line segments. The first is acceleration from rest to 6 ms^{-1}, the second is motion at a uniform velocity and the third is deceleration back to rest. Find

a) T, the time that it takes to accelerate if the initial acceleration is 0.2 ms^{-1}

b) the total distance travelled by the elevator

c) the average speed of the elevator.

8. 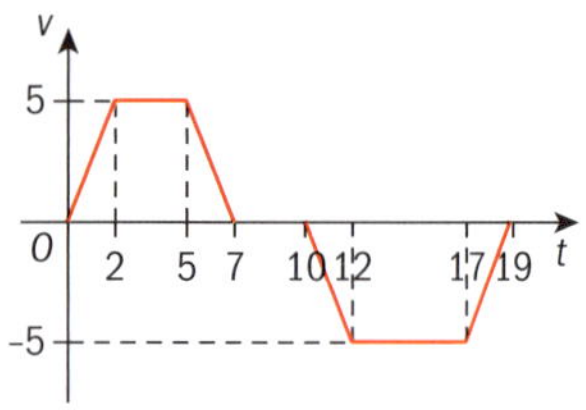

The velocity–time graph shows an elevator travelling in a building. There are seven stages in its journey. It accelerates from rest to 5 ms^{-1}, and then travels at constant velocity before decelerating back to rest. It remains stationary before moving downwards, again accelerating, moving with constant velocity and decelerating. The times taken for each of these stages are shown on the graph. Calculate

a) the rates at which the elevator accelerates and decelerates

b) the distance travelled moving upwards

c) the distance travelled moving downwards.

d) If each floor in the building measures 2.5 m and the elevator starts on the 8th floor, on which floor does it first stop, and on which floor is it when it stops after 19 seconds?

9. 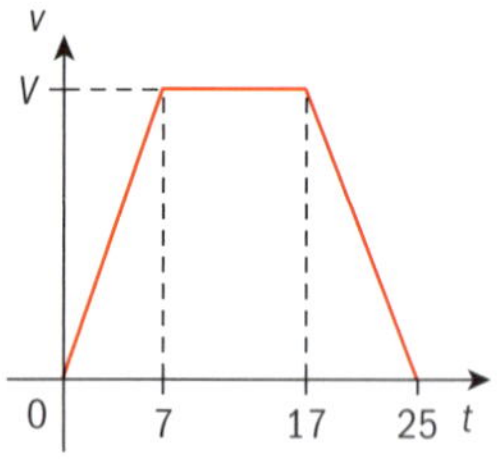

Use the graph. Given that the total displacement is 420 m, find

a) the value of V

b) the acceleration at time $t = 5$

c) the acceleration at time $t = 20$

d) the times at which the speed is 6 ms^{-1}.

10. Use the graph. If the total displacement is 380 m, find

a) the value of V

b) the value of the deceleration when $t = 8$ and when $t = 30$

c) the average speed for the whole journey

d) the time taken to travel the first 180 m of the journey.

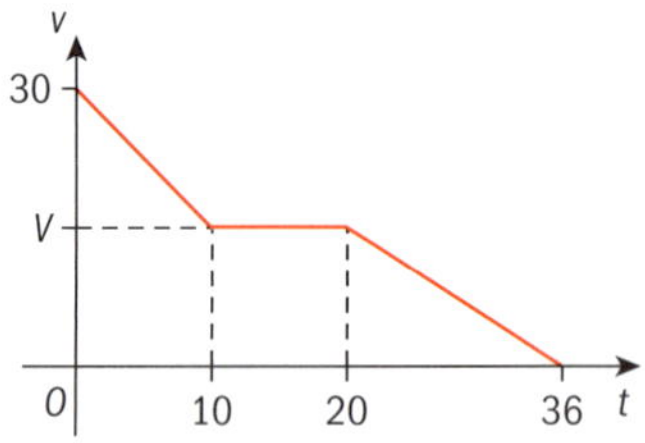

11.

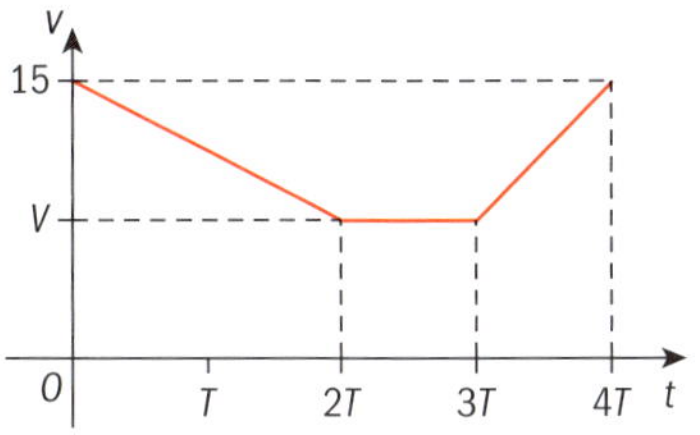

Use the graph. Given that the total distance travelled is 70 m and that the initial deceleration is 2.5 ms^{-2}, find

a) the possible values of V

b) the possible values of T.

12.

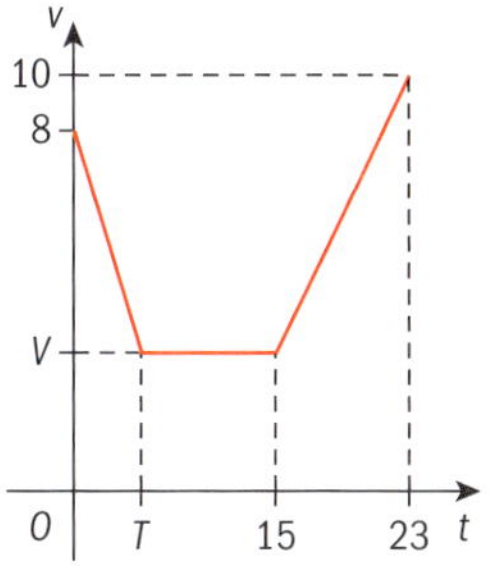

Use the graph. If the total displacement is 138 m and the initial deceleration is 1 ms^{-2}, find

a) the value of V

b) the value of T.

13. A ball is thrown upwards at a speed of 20 ms^{-1} from a tower that is 25 m high. It goes up into the air and then falls all the way to the ground without hitting the tower. When it reaches the ground, it bounces back up at half the speed it hit the ground with. It comes to rest when it hits the ground for the second time. Times are shown on the velocity–time graph. If the acceleration due to gravity is -10 ms^{-2}, find

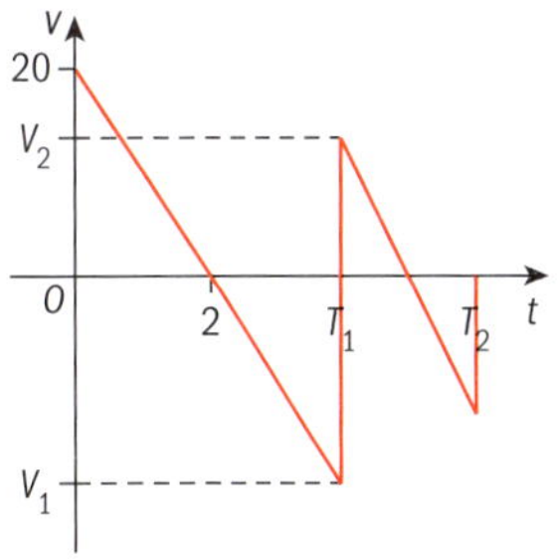

a) the greatest height it reaches above the ground

b) the time it takes to reach the ground

c) its velocity V_1 when it first hits the ground

d) the velocity V_2 when it rebounds

e) the greatest height it reaches above the ground after it bounces

f) the total time T taken for the ball to come to rest.

Summary exercise 1

1. In a game of pool, a ball is hit towards the cushion at a velocity of $10\,\text{ms}^{-1}$. When it hits the cushion, it changes direction, returning at a velocity of $8\,\text{ms}^{-1}$. It takes 0.04 s for the ball to reach the cushion.

   a) Find the distance that the ball travels between the starting point and the cushion.

   b) Sketch a displacement–time graph showing the motion of the ball until the time that it returns to its starting position.

2. Two trains are at opposite ends of a straight track, which is 800 m long and runs from A to B. The train at A departs at time $t = 0$ at a speed of $20\,\text{ms}^{-1}$. At time $t = 5$, the train at B departs at a speed of $32\,\text{ms}^{-1}$. When it is halfway to A, the second train stops for 5 s, before proceeding at $20\,\text{ms}^{-1}$.

   a) Sketch, on the same diagram, a displacement–time graph showing the motion of the two trains.

   b) Where and at what time do the two trains pass each other?

   c) Which train arrives at its destination first?

3. 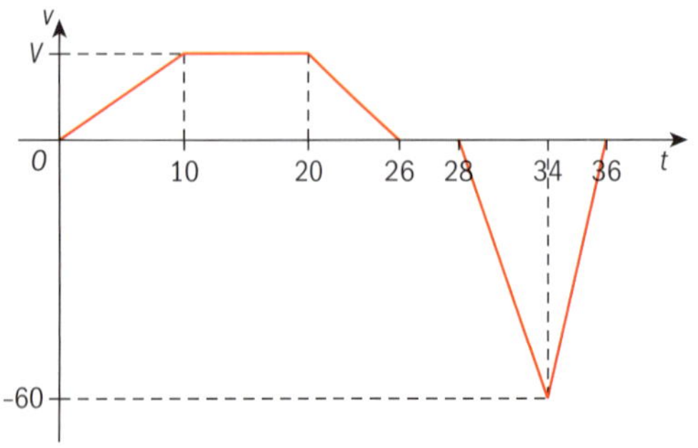

   A roller coaster in an amusement park is known as "The Long Drop". The ride takes just over half a minute from beginning to end. The car starts at ground level and is carried upward, accelerating to a velocity V, then decelerating and coming to a stop at the top of the ride. At the top of the track, it waits for 2 s before dropping back to the ground. After 6 s it decelerates rapidly, coming to a halt at ground level after a further 2 s. It reaches a maximum speed of $60\,\text{ms}^{-1}$ on its descent. Calculate

   a) the distance that the car falls on its descent

   b) the maximum velocity, V, that it reaches on its ascent

   c) the value of the deceleration in the final phase before the car comes to a halt

   d) the distance that the car drops before it begins to decelerate.

4. 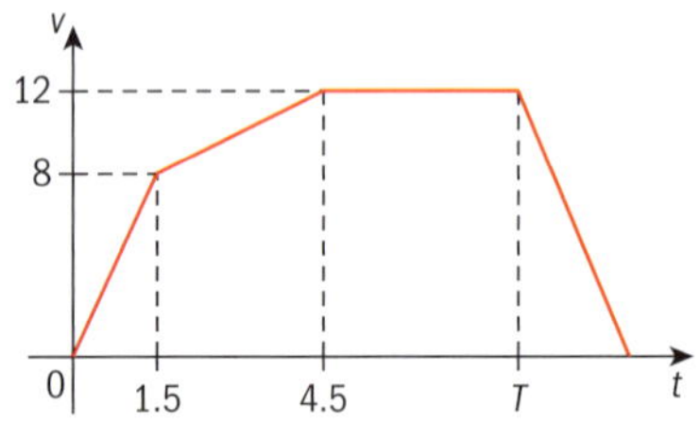

   A 100 m sprinter accelerates to $8\,\text{ms}^{-1}$ in 1.5 s. He then accelerates to $12\,\text{ms}^{-1}$ in the next 3 s and runs at constant speed for the remainder of the race. He completes the race in T s. After he passes the finish, he decelerates and stops further along the track. Find

   a) the time he takes to complete the race

   b) his rate of deceleration at the end of the race if he runs a further 20 m before coming to a stop.

5. A train starts from rest at a station A and travels in a straight line to station B, where it comes to rest. The train moves with constant acceleration $0.025\,\text{ms}^{-2}$ for the first 600 s, with constant speed for the next 2600 s, and finally with constant deceleration $0.0375\,\text{ms}^{-2}$.

   i) Find the total time taken for the train to travel from A to B. [4]

   ii) Sketch the velocity–time graph for the journey and find the distance AB. [3]

   iii) The speed of the train t seconds after leaving A is $7.5\,\text{ms}^{-1}$. State the possible values of t. [1]

Cambridge International AS and A Level Mathematics 9709, Paper 41 Q5 May/June 2011

6. 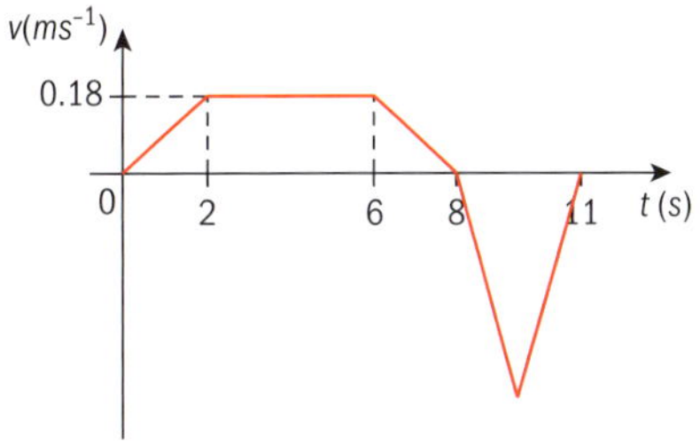

The diagram shows the velocity–time graph for the motion of a machine's cutting tool. The graph consists of five straight line segments. The tool moves forward for 8 s while cutting and then takes 3 s to return to its starting position. Find

   i) the acceleration of the tool during the first 2 s of the motion, [1]

   ii) the distance the tool moves forward while cutting, [2]

   iii) the greatest speed of the tool during the return to its starting position. [2]

Cambridge International AS and A Level Mathematics 9709, Paper 42 Q2 May/June 2010

Chapter summary

Displacement–time graphs

- A displacement–time graph is used to show the motion of a **particle**, in one dimension, along a straight line.
- Velocity = gradient of displacement–time graph
- Displacement is often denoted by s.

Velocity–time graph

A velocity-time graph is also used to show the motion of a particle in one dimension, along a straight line.

- Acceleration = gradient of velocity–time graph
- Displacement = area under velocity–time graph
   - For constant velocity, displacement is found by calculating the area of the rectangle on a velocity–time graph.
- Average speed $= \dfrac{\text{total displacement}}{\text{time taken}}$

It is a common error to use the incorrect formula for average speed so make sure to learn this formula.

2 Constant acceleration formulae

Sprinters at the start of a race increase their speed at a constant rate. They are accelerating **uniformly** for the first few seconds of the race until they reach top speed. In the 100 m race, the track is a straight line, and sprinters typically reach their top speed after 50–60 m. There are many situations in which the acceleration formulae we are about to study can be used to model motion in a straight line with constant (uniform) acceleration. For example, the acceleration due to gravity is constant, and these formulae are used in a huge number of situations involving falling bodies.

Objectives

- Understand the concepts of distance and speed as scalar quantities, and of displacement, velocity and acceleration as vector quantities (in one dimension only).
- Use appropriate formulae for motion with constant acceleration in a straight line.

Before you start

You should know how to:

1. Substitute values into a formula.

e.g. If $a = 3$, $b = 4$ and $c = -5$, then the value of $u = a^2 + 2bc$ is

$u = 3^2 + 2 \times 4 \times (-5) = -31$

2. Solve linear equations.

e.g. Solve $8 = -2 + 4t$:

$8 + 2 = 4t$

$10 = 4t$

$t = 2.5$

3. Solve quadratic equations by factorizing or using the quadratic formula.

e.g. Solve the following quadratic equations for t.

a) $t^2 - 7t + 10 = 0$

Factorize the quadratic to get $(t - 2)(t - 5) = 0$, giving $t = 2$ or $t = 5$.

Skills check:

1. Given that $a = 3$, $b = 4$ and $c = -2$, evaluate

a) $2a^2 - b$ **b)** $a(b - c)$ **c)** $\frac{b^2 - c^2}{2a}$

2. Solve these linear equations:

a) $2u + 7 = 15$

b) $9 - 3a = 1$

c) $4t - 11 = -5$

3. Solve these quadratic equations, giving answers to 3 s.f. where appropriate.

a) $t^2 - 8t + 12 = 0$

b) $2t^2 - 3t - 1 = 0$

c) $3t^2 + 2t = 4$

b) $t^2 + 2t - 6 = 0$

Use the quadratic formula

$t = \dfrac{-b \pm \sqrt{b^2 - 4ac}}{2a}$ to get

$t = \dfrac{-2 \pm \sqrt{2^2 - 4 \times 1 \times (-6)}}{2 \times 1}$, leading

to $t = 1.65$ (to 3 s.f.) or $t = -3.65$ (to 3 s.f.).

2.1 Constant acceleration formulae

When the motion of a body is being considered, the conventional variables that we use are

s = displacement

u = initial velocity

v = final velocity

a = acceleration

t = time.

When working with variables such as these, it is important to make the distinction between **scalar** and **vector** quantities:

- Scalar quantities have magnitude (size) only.
- Vector quantities have magnitude as well as direction.

Excluding time, the characteristics of motion are vector quantities, where the direction is equally as important as the magnitude.

- **Displacement** is a vector quantity, which gives the position of a body relative to an origin.
- **Distance** is a scalar quantity that states how far the body has travelled.
- **Velocity** is a vector quantity, which tells us how fast the body is moving and in what direction.
- **Speed** is a scalar quantity, which tells us how fast the body is moving only. It is the magnitude of the velocity.

In Chapter 1, a number of relationships were seen between variables. You looked at graphs which represented the displacement, velocities and acceleration of bodies in motion. There are several simple formulae we use when dealing with constant acceleration.

Constant acceleration is also known as **uniform** acceleration.

Consider the velocity–time graph on the right showing the motion of a body with initial velocity u and final velocity v after t seconds have elapsed.

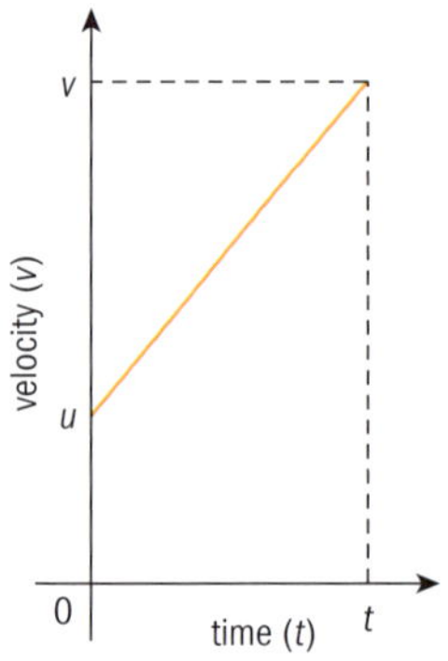

The gradient of the line is calculated from the expression

$$\frac{v-u}{t}$$

Since the gradient of this line is equal to the value of the acceleration, a, then

$$a = \frac{v-u}{t}$$

This can be rewritten to make v the subject, as

$$\boxed{v = u + at}$$ 1

This formula can be used in problems where 3 of the 4 quantities are known. However this is not always the case. Consider the area of the trapezium under the velocity–time graph. We know from Chapter 1 that this area is equal to the displacement, s, of the body, and so we obtain the formula

$$\boxed{s = \frac{1}{2}(u + v)t}$$ 2

From the graph, we have derived two important formulae. If we eliminate v from equations (1) and (2), we obtain

$$s = \frac{1}{2}(u + (u + at))t$$

which simplifies to

$$\boxed{s = ut + \frac{1}{2}at^2}$$ 3

If we eliminate t from equations (1) and (2), we obtain

$$s = \frac{1}{2}(u+v)\times\frac{v-u}{a}$$

which simplifies and rearranges to

$$\boxed{v^2 = u^2 + 2as}$$ 4

Alexandra

~~07707 653456~~

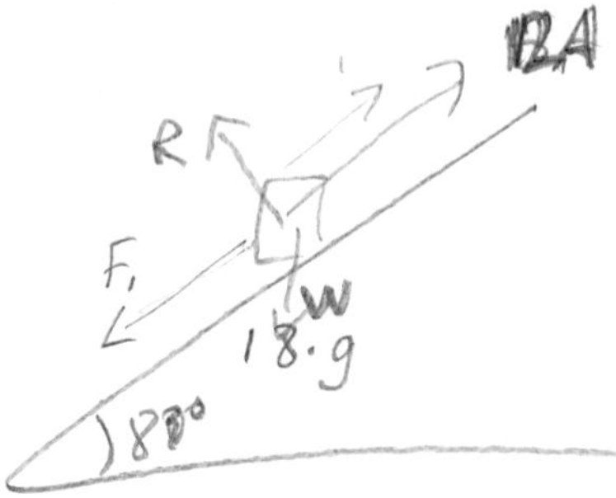

$F_1 = \mu R = \mu W \sin 82$

$F_{(max)} = \mu \times W \sin 82$ $F_{(max)} = \mu W \sin 82$

$F_2 + 5A = W \sin 82$

$F_1 + W \sin 82 = 12A$

$F_1 = 12A - W \sin 82$

$F_2 = W \sin 82 - 5A$

$F_1 + F_2 = 7A$ $A = \frac{F_1 + F_2}{7}$

$F_2 + 5\left(\frac{F_1 + F_2}{7}\right) = W \sin 82$

$F_1 + W \sin 82 = 12\left(\frac{F_1 + F_2}{7}\right)$

These four formulae, known as the **constant acceleration formulae**, can be used to solve problems provided that we can make the modelling assumption that acceleration is uniform.

Note that when acceleration is not uniform, these formulae are not valid and should not be used. This will be discussed later in the book.

These equations are also commonly referred to as the *suvat* equations.

Example 1

A car travelling at 6 ms^{-1}, accelerates at 2 ms^{-2}. Find its speed 8 seconds later.

s is not required

$u = 6$

$v = ?$

$a = 2$

$t = 8$

The formula that involves u, v, a, t is $v = u + at$

So $v = 6 + 2 \times 8 = 22\,ms^{-1}$.

Example 2

A car, initially travelling at 8 ms^{-1}, accelerates at a constant rate of 3 ms^{-2} until it is travelling at 17 ms^{-1}. Find

a) the distance travelled while accelerating

b) the time taken to accelerate.

a) $s = ?$

$u = 8$

$v = 17$

$a = 3$

t is not required

Use $v^2 = u^2 + 2as$: $17^2 = 8^2 + 2 \times 3 \times s$

So $s = 37.5$ m

b) s is not required

$u = 8$

$v = 17$

$a = 3$

$t = ?$

Use $v = u + at$: $17 = 8 + 3 \times t$

So $t = 3$ s

Example 3

According to driving guidelines the minimum braking distance for a car travelling at 20 ms^{-1} is 30 m. Find

a) the deceleration of the car

b) the time it would take for the car to stop.

Braking distance is the distance a car travels from the point when its brakes are applied to the point when it comes to a complete stop.

a) $s = 30$

$u = 20$

$v = 0$

$a = ?$

t is not required

Use $v^2 = u^2 + 2as$:

$0^2 = 20^2 + 2 \times a \times 30$

So $a = -6\frac{2}{3}\,ms^{-2}$

Hence the deceleration is $6\frac{2}{3}\,ms^{-2}$.

b) $s = 30$

$u = 20$

$v = 0$

a is not required

$t = ?$

Use $s = \frac{1}{2}(u + v)t$:

$30 = \frac{1}{2}(20 + 0)t$

So $t = 3$ s

Example 4

A particle passes a point A with speed of 2 ms^{-1}. The particle is accelerating at a constant rate of 4 ms^{-2}. Find the time taken for the particle to be 17.5 m from A.

$s = 17.5$

$u = 2$

v is not required

$a = 4$

$t = ?$

The formula that involves s, u, a, t is $s = ut + \frac{1}{2}at^2$:

$17.5 = 2t + \frac{1}{2} \times 4 \times t^2$ (note that this is a quadratic equation in t)

Solving gives $t = 2.5$ or $t = -3.5$.

In the context of the problem a negative time is meaningless, so $t = 2.5$ s.

Note: In the Examples 1 – 4, all of the bodies do have size, but the modelling assumptions we make are for all bodies to be particles so that air resistance can be ignored.

Exercise 2.1

In questions **1** to **10**, a particle is moving with constant acceleration $a\,\text{ms}^{-2}$ along a straight line. The velocity at the point O is $u\,\text{ms}^{-1}$ and t seconds later the velocity is $v\,\text{ms}^{-1}$. The displacement from O at time t is s metres.

1. Find s when $u = 0$, $a = 4$, $t = 8$.

2. Find s when $u = 3$, $v = 5$, $t = 10$.

3. Find v when $u = 3$, $a = 2$, $t = 6$.

4. Find u when $s = 28$, $a = 1$, $t = 4$.

5. Find a when $s = 500$, $u = 1$, $t = 10$.

6. Find t when $u = 3$, $v = 7$, $a = 0.2$.

7. Find t when $u = 5$, $a = 1$, $s = 12$.

8. Find s when $u = 2$, $v = 10$, $a = 4$.

9. Find u when $s = 50$, $v = 6$, $t = 10$.

10. Find v when $s = 132$, $a = 2$, $t = 12$.

11. A train leaves a station and accelerates uniformly at a rate of $0.4\,\text{ms}^{-2}$. The train is in motion for 50 seconds.

a) Find how far from the station the train is at this time.

b) Find the speed of the train at this time.

12. A car passes a point O with speed $4\,\text{ms}^{-1}$. It accelerates at a constant rate of $2.5\,\text{ms}^{-2}$ to a speed of $9\,\text{ms}^{-1}$. It then decelerates at a rate of $2\,\text{ms}^{-2}$, reducing its speed to $3\,\text{ms}^{-1}$.

a) Find the distance that the car travels while accelerating. Hence find the total distance travelled by the car.

b) Find the total time the car takes to travel this distance.

13. A particle moves along a straight line AB with constant acceleration $0.5\,\text{ms}^{-2}$. If AB is 15 m and it takes 3 seconds to travel from A to B, find the velocity of the particle at A.

14. A man drives a car with constant acceleration of $4\,\text{ms}^{-2}$. After 2 seconds of accelerating he sees a set of traffic lights and slows down with a deceleration of $1.5\,\text{ms}^{-2}$. Given that the initial velocity of the car is $1\,\text{ms}^{-1}$ and the car stops at the traffic lights, find the distance between the traffic lights and the point where he starts to accelerate.

15. An elevator ascends from rest with an acceleration of $0.6\,\text{ms}^{-2}$, before slowing down with deceleration of $0.8\,\text{ms}^{-2}$ for the next stop. The total time taken is 10 seconds. Find the distance between the stops.

16. A car travels in a straight line with constant acceleration $a\,\text{ms}^{-2}$. It passes through points A, B and C, in this order, with speeds $5\,\text{ms}^{-1}$, $7\,\text{ms}^{-1}$, and $8\,\text{ms}^{-1}$ respectively. The distances AB and BC are d_1 m and d_2 m respectively.

i) Write down an equation connecting

a) d_1 and a, [2]

b) d_2 and a. [2]

ii) Hence find d_1 in terms of d_2.

Cambridge International AS and A Level Mathematics 9709, Paper 4 Q1 October/November 2005

2.2 Vertical motion

It was the work of **Isaac Newton (1642 – 1727)** that made us realize that objects in **free fall** accelerated towards the ground at a constant rate.

The constant acceleration formulae developed in section 2.1 may be used when considering the motion of bodies falling under gravity. In such cases, the acceleration of the body is widely used as $9.8\,\text{ms}^{-2}$, but often when you are set a question, it will state that an approximation of $10\,\text{ms}^{-2}$ is to be used. This value is usually referred to as g, the acceleration due to gravity.

Note: Questions in Cambridge examinations will always use $g = 10\,\text{ms}^{-2}$

Sign convention

When working through any example, care is needed to ensure that the directions of vectors involved are consistent. One way of doing this is to adopt a sign convention, where the positive direction is designated before assigning values to the *suvat* variables.

Example 5

A ball is thrown vertically downwards from the top of a building with speed $2.5\,\text{ms}^{-1}$. If the height of the building is 20 m, find

a) the speed with which the ball hits the ground

b) the time taken for the ball to reach the ground.

a) If we adopt the convention that positive is downwards for this question, then

$s = 20$

$u = 2.5$

$v = ?$

$a = g = 10$

t is not required

Use $v^2 = u^2 + 2as$:

$v^2 = 2.5^2 + 2 \times 10 \times 20$

So $v = 20.15564437 \approx 20.2\,\text{ms}^{-1}$ (to 3 s.f.)

b) $s = 20$

$u = 2.5$

$v = 20.15564437$

$a = g = 10$

$t = ?$

Use $v = u + at$:

$20.15564437 = 2.5 + 10t$

So $t = 1.76554437 \approx 1.78t\,\text{s}$ (to 3 s.f.)

Example 6

A ball is thrown vertically upwards from ground level with speed $15\,\text{ms}^{-1}$. Find

a) the greatest height reached by the ball

b) the time taken for the ball to return to ground level.

a) If we adopt the convention that positive is upwards for this question then

$s = ?$

$u = 15$

$v = 0$

$a = -g = -10$

t is not required

Use $v^2 = u^2 + 2as$:

$0^2 = 15^2 + 2 \times (-10) \times s$

So $s = 11.25\,\text{m}$

b) $s = 0$

$u = 15$

v is not required

$a = -g = -10$

$t = ?$

Use $s = ut + \frac{1}{2}at^2$:

$0 = 15t + \frac{1}{2} \times (-10) \times t^2$

So $t = 0\,\text{s}$ or $t = 3\,\text{s}$

Hence, the time taken to return to ground level is 3 seconds.

Note that in Examples 5 and 6, we use the model of ignoring air resistance, so that the acceleration is then constant and equal to the acceleration due to gravity. This also means that the motion in Example 6 is symmetric. So, for part **(b)**, we could have worked out the time to the highest point (1.5 seconds from using $v = u + at$) which is the same as the time from the highest point back to ground level, hence time required is double the time to the highest point. You can use this method to solve questions of the same type as in Examples 5 and 6.

Exercise 2.2

In this exercise, take g as $10\,\text{ms}^{-2}$ and give answers correct to three significant figures where appropriate.

1. A book falls from a shelf 1.8 m above the floor. Find the speed with which the book strikes the floor.

2. A stone is dropped from 48 m above the ground. Find the time it takes for the stone to reach the ground.

3. A stone is dropped from the top of a cliff and falls to ground level. If the stone hits the ground at $18\,\text{ms}^{-1}$, find the height of the cliff.

4. A ball is thrown vertically upwards with speed 25 ms^{-1} and travels freely under gravity. Find the velocity of the ball after 2 seconds, and the distance the ball has travelled from the start at this time.

5. For the ball in question **4**, find the maximum height reached by the ball, and the total time taken to return to its starting position.

6. A stone is thrown vertically upwards with speed 12 ms^{-1}, from 5 m above horizontal ground. Find

   a) the speed with which the stone hits the ground

   b) the time taken for the stone to hit the ground.

7. A ball is dropped from a height of 30 m. Find

   a) the time taken for the ball to reach the ground

   b) the speed at which the ball hits the ground.

8. A ball is thrown vertically upwards with initial speed of 8 ms^{-1} from a height of 1 m above level ground. Find

   a) the time when the speed of the ball is zero

   b) the greatest height reached by the ball

   c) the speed of the ball when it hits the ground.

9. A boy drops a ball from rest from the top of a building. At the same time, his friend throws a ball vertically upwards from the base of the building with speed 30 ms^{-1}. The two balls collide after 1.8 s. Find the distance from the ground to the top of the building.

10. A ball is thrown vertically upwards from a point O with speed 30 ms^{-1}.

    a) Find the time that the ball is a height of 25 m above O

    i) for the first time

    ii) for the second time.

    b) Find the total time that the height of the ball above O is at least 25 m.

11. A particle is projected vertically upwards from a fixed point O. The speed of projection is u ms^{-1}. The particle returns to O 4 seconds later. Find

    a) the value of u

    b) the greatest height reached by the particle

    c) the total time for which the particle is at a height greater than half its greatest height.

12. An object is projected vertically upwards with speed $9\,\text{ms}^{-1}$. Calculate

  a) the speed of the object when it is 2.4 m above the point of projection

  b) the greatest height of the object above the point of projection

  c) the time after projection when the object is travelling downwards with speed $4.6\,\text{ms}^{-1}$.

13. A particle P is projected vertically upwards from horizontal ground with speed $8\,\text{ms}^{-1}$.

  a) Show that the greatest height above ground reached by P is 3.2 m.

  A particle Q is projected vertically from a point 1.4 m above the ground, with speed $u\,\text{ms}^{-1}$. The greatest height above the ground reached by Q is also 3.2 m.

  b) Find the value of u.

  c) Find the speed and direction of both particles when P and Q are at the same height.

14. A stone is released from rest and falls freely under gravity. Find

  (i) the speed of the stone after 2 s, [1]

  (ii) the time taken for the stone to fall a distance of 45 m from its initial position, [2]

  (iii) the distance fallen by the stone from the instant when its speed is $30\,\text{ms}^{-1}$ to the instant when its speed is $40\,\text{ms}^{-1}$. [2]

Cambridge AS and A Level
Mathematics 9709 Q2 Paper 4
October/November 2003

Summary exercise 2

1. A ball is dropped onto level ground from a height of 20 m.

   a) Calculate the time taken for the ball to reach the ground.

   The ball rebounds (bounces back) with half the speed it strikes the ground.

   b) Calculate the time taken for the ball to reach the ground a second time after the initial bounce.

2. A rocket is travelling with a velocity of 80 ms^{-1}. The engines are switched on for 8 seconds and the rocket accelerates uniformly at 30 ms^{-2}.

   a) Calculate the speed of the rocket immediately after the engines are turned off.

   b) Calculate the distance travelled by the rocket while accelerating.

EXAM-STYLE QUESTION

3. A top sprinter in the 100 m race will accelerate at a rate of 5.5 ms^{-2} in the first 2 seconds of the race.

   a) Find how far the sprinter runs while accelerating.

   b) Assuming that the sprinter runs the rest of the race at the speed he attained after 2 seconds, find the time he takes to complete the race.

4. A ball is thrown vertically upwards from the top of a cliff, which is 80 m high. The initial speed of the ball is 25 ms^{-1}. Find the time taken to reach the bottom of the cliff and the speed of the ball at that instant.

EXAM-STYLE QUESTIONS

5. A ball is thrown vertically upwards with a speed of 45 ms^{-1}. Find the length of time for which the ball is at least 25 m above the point of release.

6. A car and a lorry are initially at rest side by side. The lorry moves off with a uniform acceleration of 0.6 ms^{-2}. After 10 seconds, the car moves off with uniform acceleration of 1.6 ms^{-2}. Find how long the lorry has been in motion when it is overtaken by the car, and find the distance travelled by the lorry in this time.

7. A car accelerates uniformly from rest at 1.2 ms^{-2}, and immediately begins to decelerate to a stop at a uniform rate of 1.6 ms^{-2}. The total distance covered is 1600 m. Find the total time for the journey, and the greatest speed attained by the car.

8. 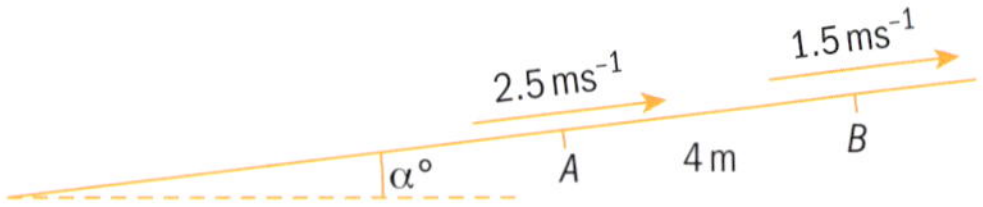

   A particle slides up a line of greatest slope of a smooth plane inclined at an angle $\alpha°$ to the horizontal. The particle passes through the points A and B with speeds 2.5 ms^{-1} and 1.5 ms^{-1} respectively. The distance AB is 4 m (see diagram). Find

   i) the deceleration of the particle, [2]

   ii) the value of α. [2]

Cambridge AS and A Level Mathematics 9709, Paper 4 Q1 May/June 2007

9. 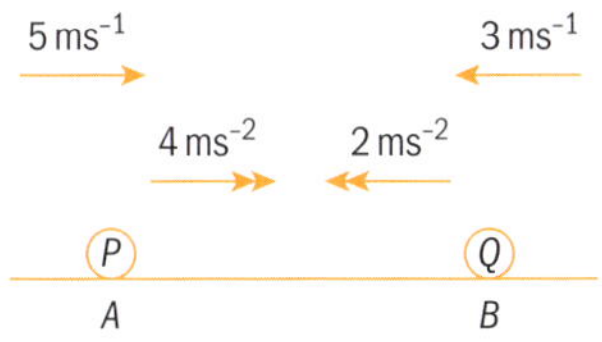

Particles P and Q start from points A and B respectively, at the same instant, and move towards each other in a horizontal straight line. The initial speeds of P and Q are $5\,\text{ms}^{-1}$ and $3\,\text{ms}^{-1}$ respectively. The accelerations of P and Q are constant and equal to $4\,\text{ms}^{-2}$ and $2\,\text{ms}^{-2}$ respectively (see diagram).

i) Find the speed of P at the instant when the speed of P is 1.8 times the speed of Q. [4]

ii) Given that $AB = 51$ m, find the time taken from the start until P and Q meet. [4]

Cambridge AS and A Level Mathematics 9709, Paper 4 Q5 October/November 2004

10. Two particles A and B are projected vertically upwards from horizontal ground at the same instant. The speeds of projection of A and B are $5\,\text{ms}^{-1}$ and $8\,\text{ms}^{-1}$ respectively. Find

i) the difference in the heights of A and B when A is at its maximum height, [4]

ii) the height of A above the ground when B is 0.9 m above A. [4]

Cambridge AS and A Level Mathematics 9709, Paper 4 Q4 October/November 2002

Chapter summary

Constant acceleration formulae

- **Displacement** is a vector quantity, which gives the position of a body relative to an origin.
- **Distance** is a scalar quantity that states how far the body has travelled.
- **Velocity** is a vector quantity, which tells us how fast the body is moving and in what direction.
- **Speed** is a scalar quantity, which tells us how fast the body is moving only. It is the magnitude of the velocity.
- $v = u + at$
- $s = \frac{1}{2}(u + v)t$
- $s = ut + \frac{1}{2}at^2$
- $v^2 = u^2 + 2as$

Note: These also know as the *suvat* equations.

Vertical motion

- The acceleration due to gravity ($g\,\text{ms}^{-2}$) acts on bodies falling vertically.
- The value of g is $9.8\,\text{ms}^{-2}$, but questions in Cambridge examinations will always use $g = 10\,\text{ms}^{-2}$.

Modelling conditions

- All bodies should be considered to be particles.
- Air resistance should be ignored.

3 Forces and resultants

Force is an influence which can cause a change in the motion of a body, or cause a stationary body to move. Intuitively, we recognise the concept of force as a "push" or "pull". We encounter forces all the time in everyday life, such as when pulling a sledge up a hill or pushing a trolley in the supermarket, and so it is an important component in most mechanical applications. Force is measured in newtons (N) and is represented by the symbol F.

Objectives

- Identify the forces acting in a given situation.
- Understand the vector nature of force, and find and use components and resultants.

Before you start

You should know how to:

1. Express line segments as vectors.

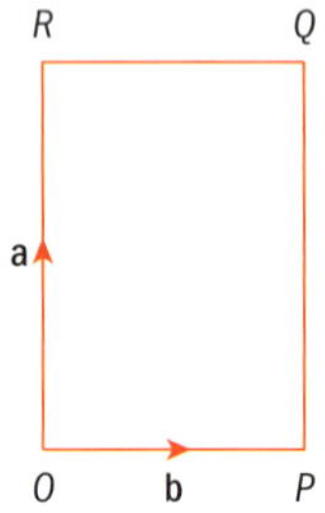

$\overrightarrow{OR} = \mathbf{a}$

$\overrightarrow{OP} = \mathbf{b}$

2. Use trigonometry to find angles and side lengths in right-angled triangles.

- $\sin\theta = \dfrac{\text{opposite}}{\text{hypotenuse}}$
- $\cos\theta = \dfrac{\text{adjacent}}{\text{hypotenuse}}$
- $\tan\theta = \dfrac{\text{opposite}}{\text{adjacent}}$

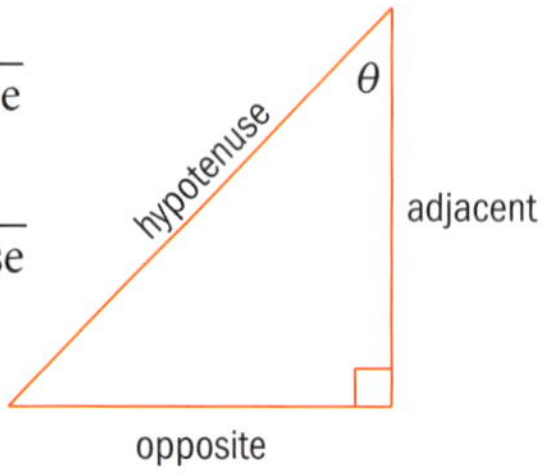

Skills check:

1. *OABC* is a trapezium. The parallel sides *OC* and *AB* are such that $OC = 3AB$.

   Given that $\overrightarrow{AB} = \mathbf{a}$, write, in terms of **a**, the vector

   a) $\overrightarrow{OC}$

   b) $\overrightarrow{CO}$

2. Raj walks from a point *A* in the direction west for 6 km and calls this point *B*. He returns to the starting point and decides to walk north for 8 km. He then walks directly to *B* from this point, covering 10 km. What angle does Raj make between the longest and shortest side?

3.1 Resultants

When two or more vectors are added together the single equivalent vector is called the **resultant vector**. We can apply the same process to forces (since they have both magnitude and direction, and can be expressed as vectors).

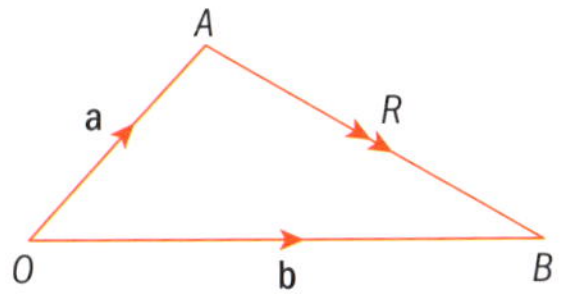

For example, if we look at the above diagram where $\overrightarrow{OA} = a$ and $\overrightarrow{OB} = b$, $\overrightarrow{AB}$ is equal to $b - a$. It is also called the resultant, R, of $\overrightarrow{OA}$ and $\overrightarrow{OB}$.

We can now extend the above idea with the introduction of a parallelogram with three vectors – $\overrightarrow{OA} = a$, $\overrightarrow{OC} = c$ and $\overrightarrow{OB} = R$, our resultant vector.

The resultant vector, R is the resultant of $\overrightarrow{OA}$ and $\overrightarrow{AB}$ and also $\overrightarrow{OC}$ and $\overrightarrow{CB}$.

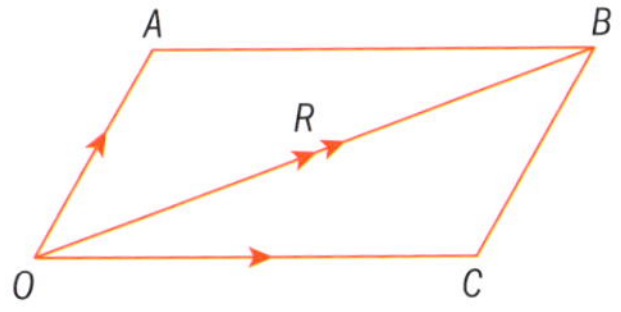

Example 1

An anchor is being pulled by two sailors as shown in the diagram. The angle between the two forces is 30°.

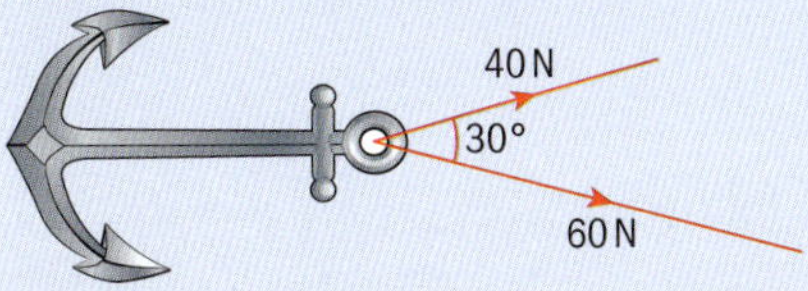

Force is measured in newtons, N.

Find the magnitude of the resultant of the two forces.

Continued on the next page

To answer this question we look towards the parallelogram of forces.

Two forces E and F are represented by the line segment AB and AD, respectively as shown in the diagram.

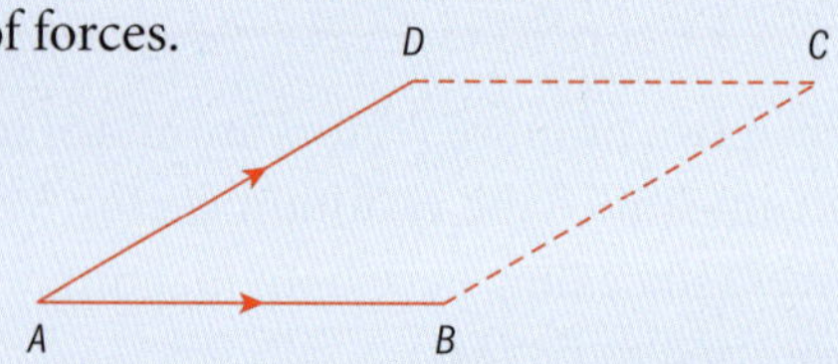

$\overrightarrow{AB} + \overrightarrow{AD} = \overrightarrow{AB} + \overrightarrow{BC} = \overrightarrow{AC}$

In other words, the resultant of the two forces E and F, which represent the line segments $\overrightarrow{AB}$ and $\overrightarrow{AD}$, respectively, can be explained by the diagonal $\overrightarrow{AC}$.

$\overrightarrow{AC}$ is the diagonal of our parallelogram $ABCD$ and is called the **parallelogram of forces.**

Going back to our question,

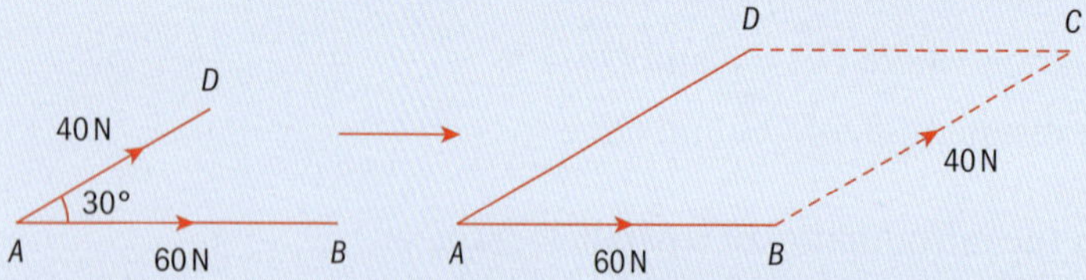

As we know, opposite angles in a parallelogram are congruent.

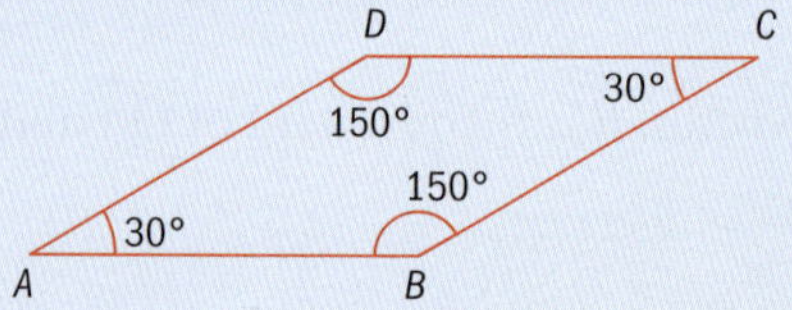

Looking back at our diagram:

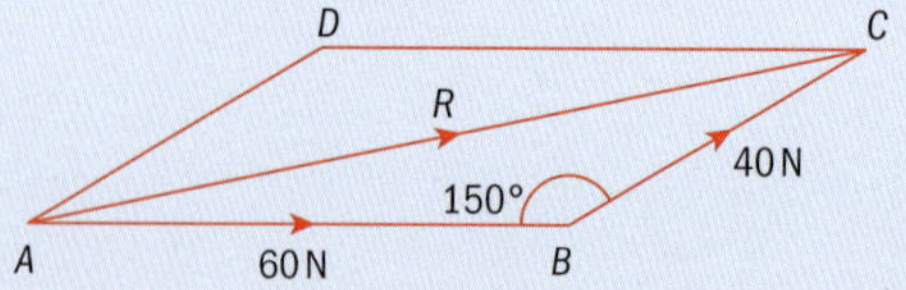

Recall the cosine rule:
$c^2 = a^2 + b^2 - 2ab \cos C$

To find the resultant $\overrightarrow{AC}$ (denoted in the diagram as R), we use the cosine rule.

$R^2 = 60^2 + 40^2 - 2 \times 60 \times 40 \times \cos 150°$

$R^2 = 9356.922\ldots$

$R = 96.73\,\text{N}$ (2 d.p.)

The resultant of the two forces is of magnitude 96.73 N.

To find the angle θ we split the parallelogram so that we have the following triangle ABC. To find θ we apply the sine rule.

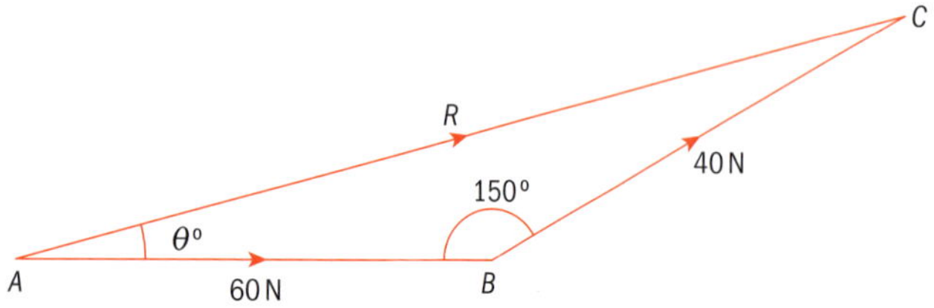

$$\frac{96.73}{\sin 150°} = \frac{40}{\sin \theta°}$$

$$\sin \theta = \frac{40 \sin 150°}{96.73}$$

$$\theta = 11.93°$$

Example 2

A boat is being towed along a canal by cables attached to two horses. The stronger horse produces a force of 300 N, and the other produces 260 N as shown in the diagram.

Find the acute angle between the two forces if their resultant has a magnitude of 540 N.

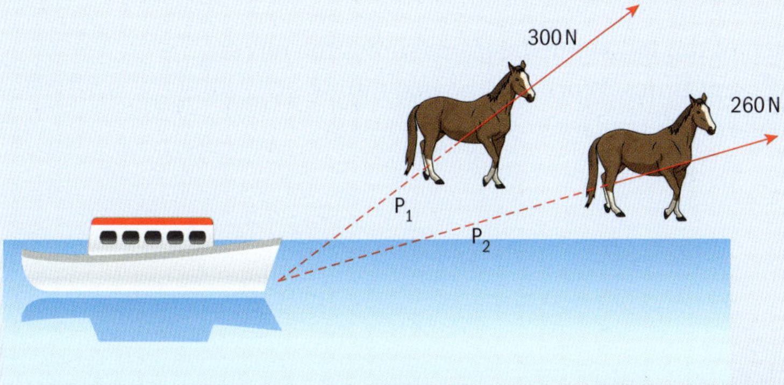

Using the parallelogram of forces:

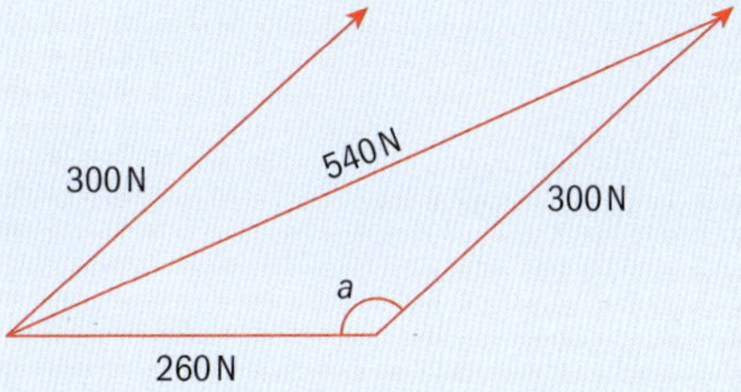

By the cosine rule,

$540^2 = 260^2 + 300^2 - 2 \times 260 \times 300 \times \cos a°$

$$\cos a° = \frac{540^2 - (260^2 + 300^2)}{-2 \times 260 \times 300}$$

$$\cos a° = -\frac{67}{78}$$

$a = (180° - 30.798°\ldots) = 149.2016°$

To find the acute angle,

$180° - a = 30.8°$

Hence, the angle between the two forces of 300 N and 260 N is 30.8° (1 d.p.)

Exercise 3.1

1. A car has broken down and it is being pulled to the garage by a team of workers. As it is hard work, a pair of workers will pull for 5 minutes each then swap over. On average the forces being applied are 10 N and 12 N, respectively. Find the magnitude and direction of the resultant forces between 10 N and 12 N if the angle between the forces is

   a) 20°   b) 45°   c) 105°

2. Two horses are pulling a cart. Each of the following diagrams shows the forces being applied by the horses at different times. Work out the magnitude of their resultant and the angle it makes with the larger of the two forces.

   a)
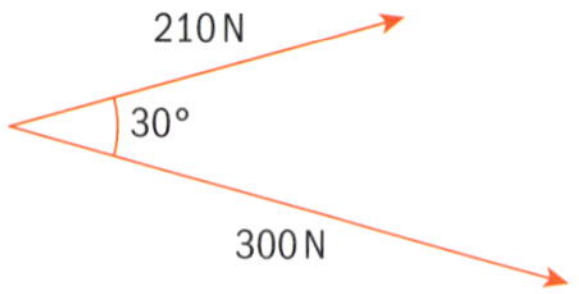

   b)
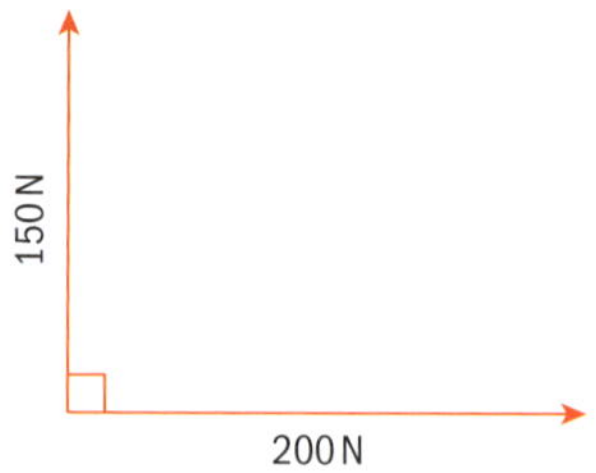

   c)
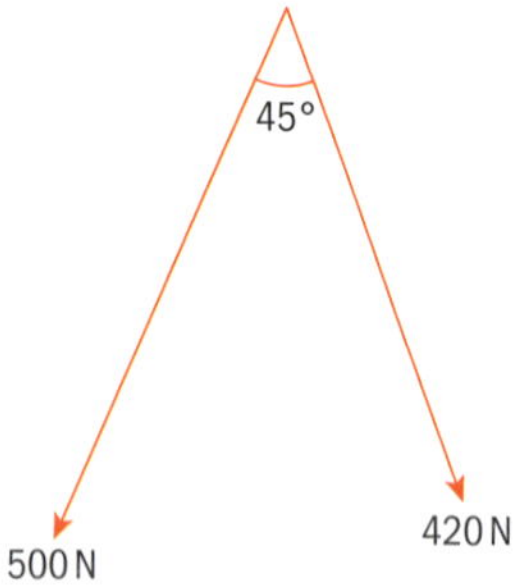

   d)
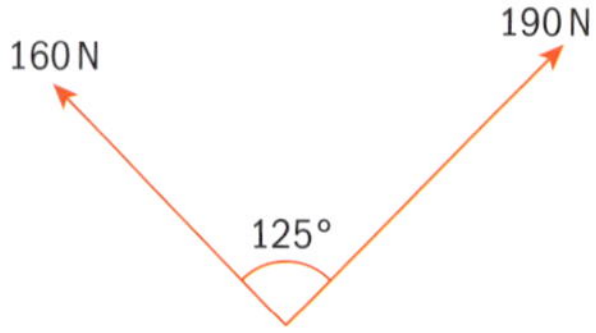

3. A tractor has broken down and is being pushed by two people. The magnitude and direction of the force 15° above the horizontal is 210 N. The magnitude and direction of the force 35° below the horizontal is 190 N, as shown in the diagram.

   Find the resultant of the two forces.

4. Two tug boats are towing a ship into harbour. One of the boats produces a pulling force of 25 000 N. The other produces 27 000 N worth of pulling force. Their resultant has a magnitude of 35 000 N. Find the angle between the two forces and the angle the resultant makes with the larger of the two forces.

5. A young boy is being pulled along on a sledge by both of his parents. In order for the sledge to move along the snow, two forces of 120 N and XN, with an angle of 65° between them, has to be maintained. If the resultant of the two forces has a magnitude of 160 N, find the value of X.

6. A caravan has just been sold at a garage, but the equipment needed to remove it has broken down. Instead, the new owner and the salesperson have decided to push the caravan to the forecourt of the garage so that the owner can connect it to his car and drive it away. If the resultant of the two forces is 620 N, find the angle x between the 450 N force and the horizontal.

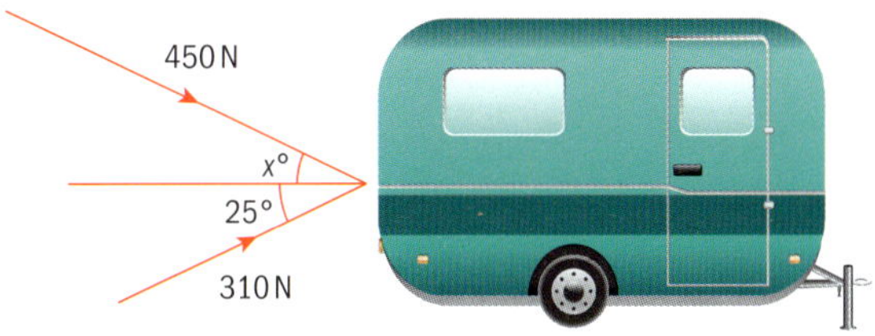

3.2 Components

Previously we have looked at combining two forces into a single force (called the resultant). We will now look at the reverse process which involves taking a single force and breaking it up into **components.** These components are also referred to as the **resolved parts** of the force. The process of finding the resolved parts of a force is called **resolving**.

When resolving forces let us consider the diagram on the right.

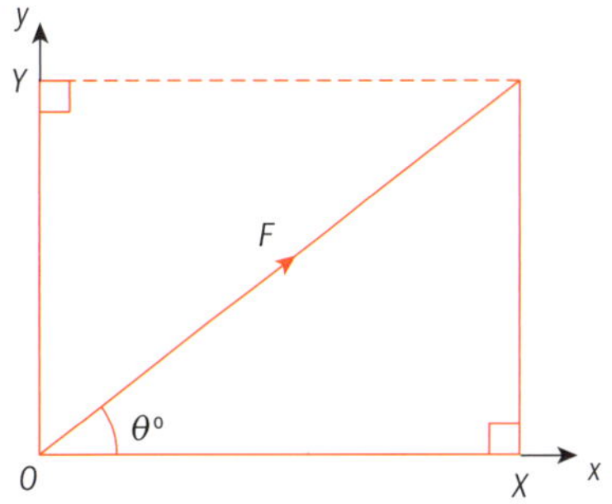

$\cos\theta = \frac{OX}{F}$ giving $OX = F\cos\theta$;

$\sin\theta = \frac{OY}{F}$ giving $OY = F\sin\theta$;

The force in the x-direction, OX is $F\cos\theta$ and in the y-direction, OY is $F\sin\theta$. A force in the x-direction is regarded as the horizontal component and we use the letter **i** to signify this. That is, the unit vector in the direction of x. Similarly, in the y-direction the letter **j** is used, the unit vector in the direction of y.

Example 3

Using the diagram, find the components of the given force in the direction of

a) the x-axis

b) the y-axis.

c) Express the force in the form $a\mathbf{i} + b\mathbf{j}$.

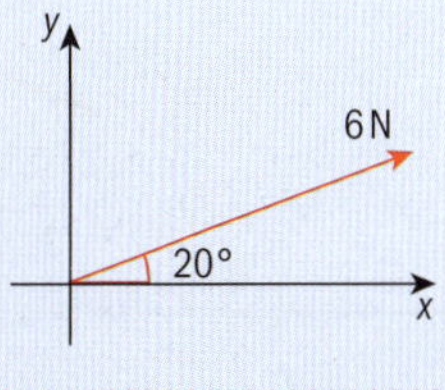

a) Along the x-axis, component $OX = 6 \times \cos 20° = 5.64$ N

b) Along the y-axis, component $OY = 6 \times \sin 20° = 2.05$ N

c) Adding the components along the x-axis and y-axis, we get: $(5.64\mathbf{i} + 2.05\mathbf{j})$ N

Exercise 3.2

For each of the following diagrams, find the components of the given force in the direction of

a) the x-axis b) the y-axis.

c) Express the force in the form $a\mathbf{i} + b\mathbf{j}$.

1.

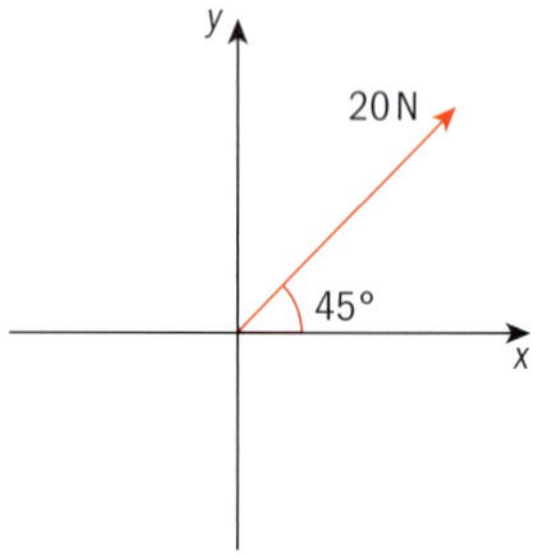

2.

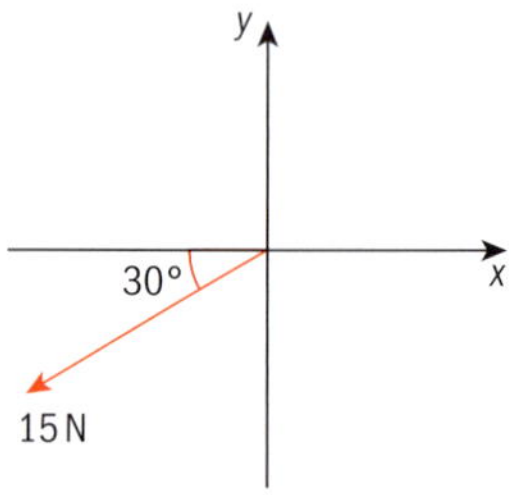

3.

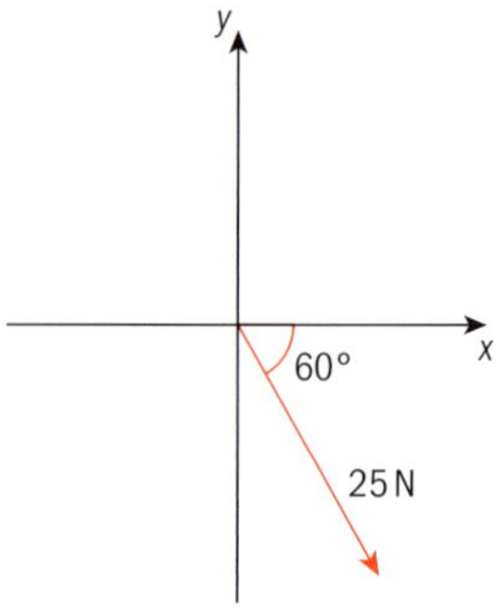

4.

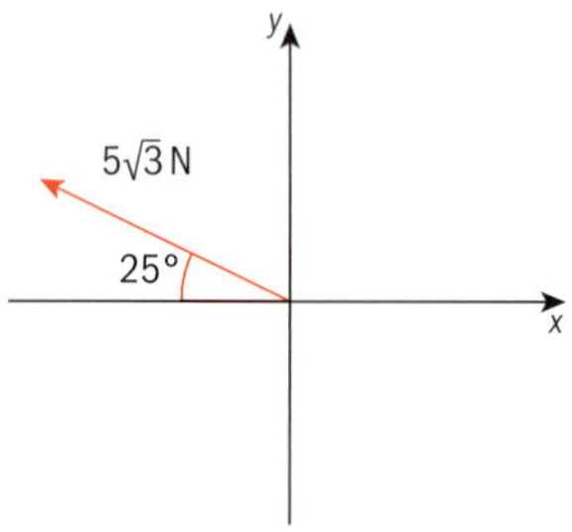

5.

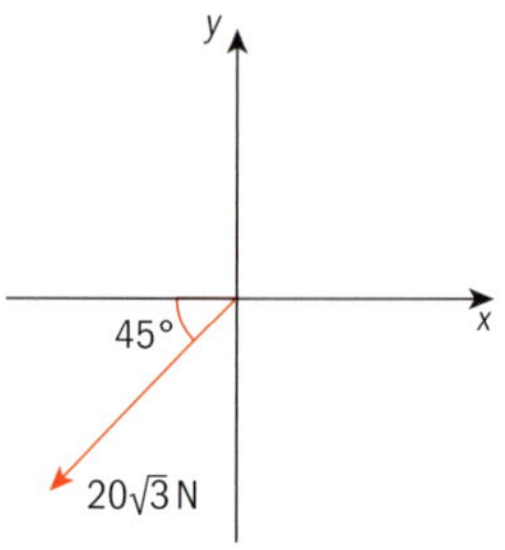

6.

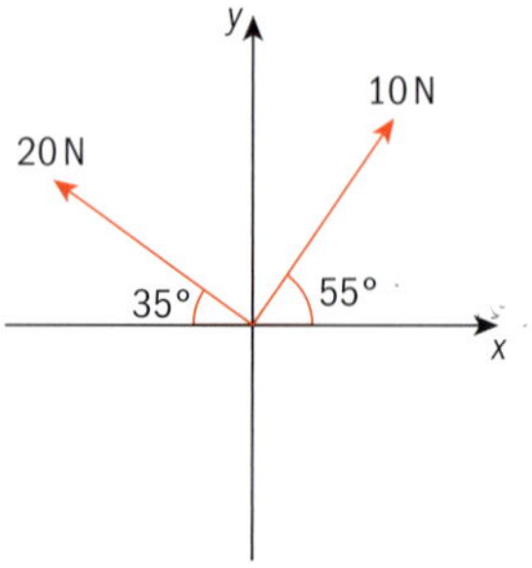

Summary exercise 3

1. For each of the forces shown in the diagram, find the components of the force in the direction of

   a) the x-axis b) the y-axis.

   c) Express the force in the form $a\mathbf{i} + b\mathbf{j}$.

i)

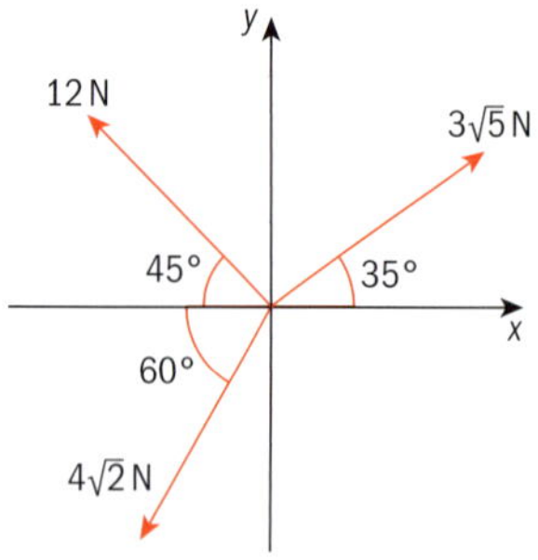

ii)

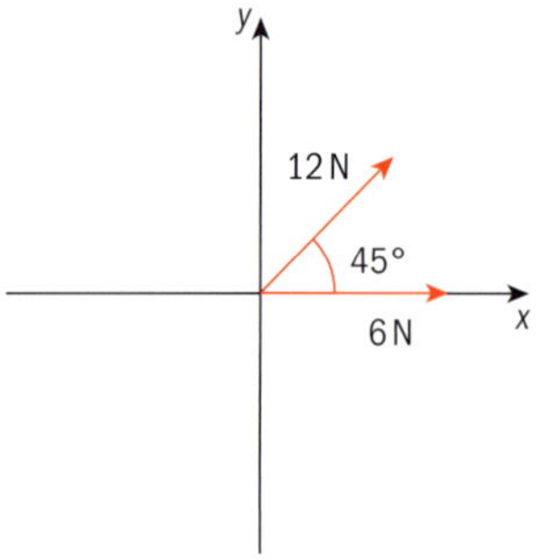

iii)

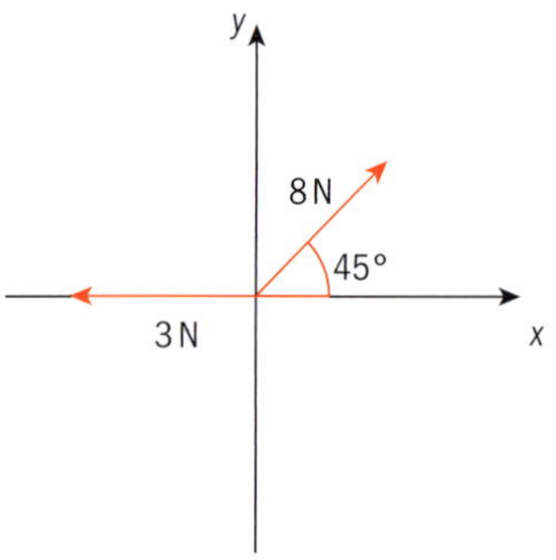

iv)

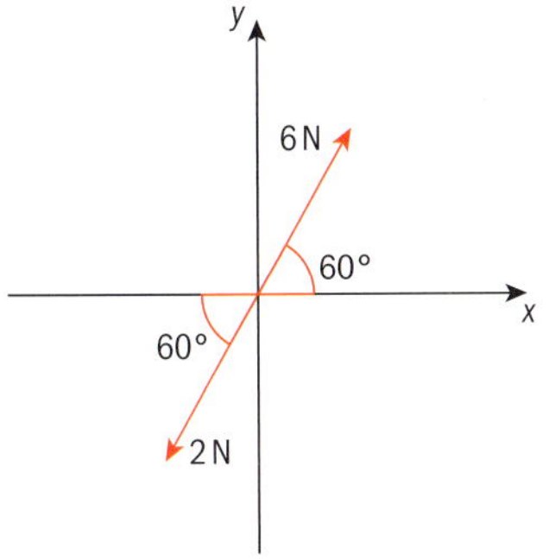

v)

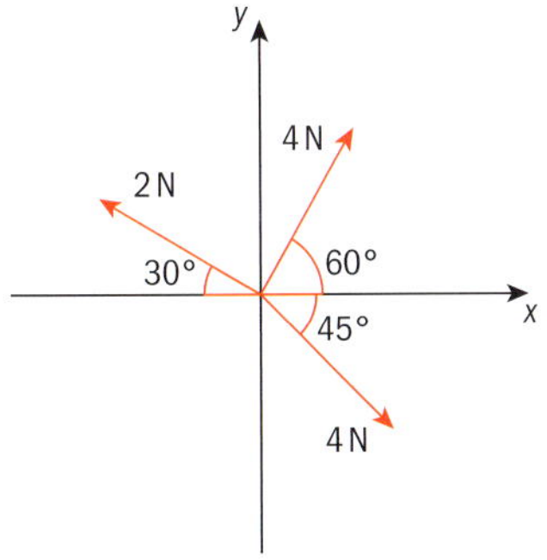

vi)

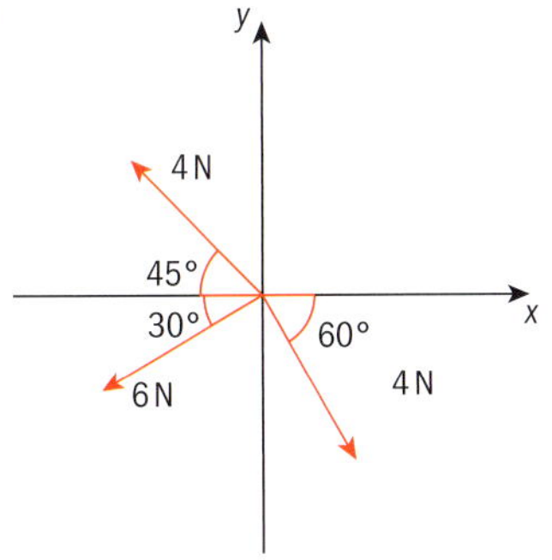

vii)

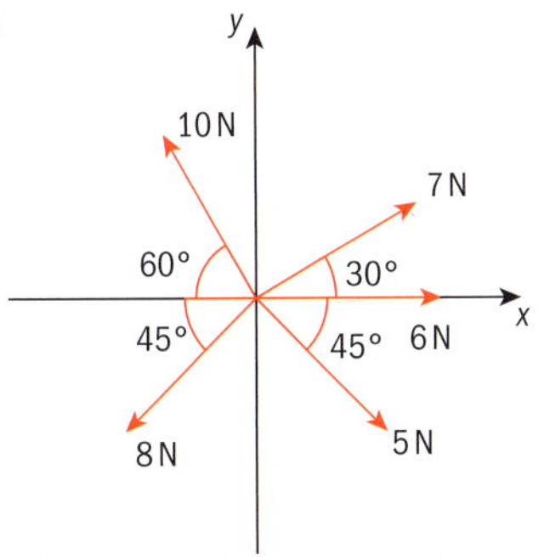

2. For each of the following system of forces, find the resultant.

a)

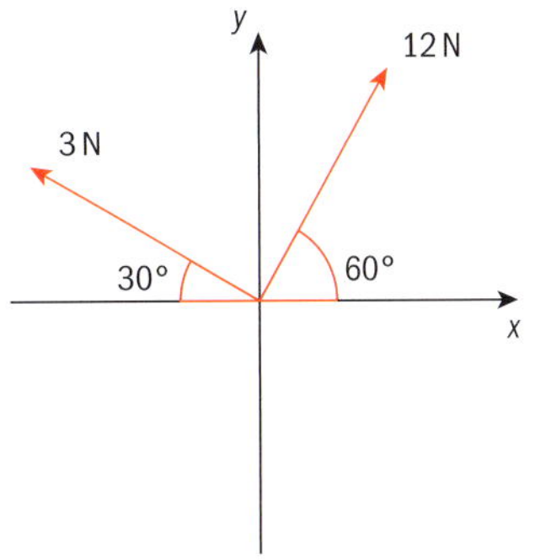

b)

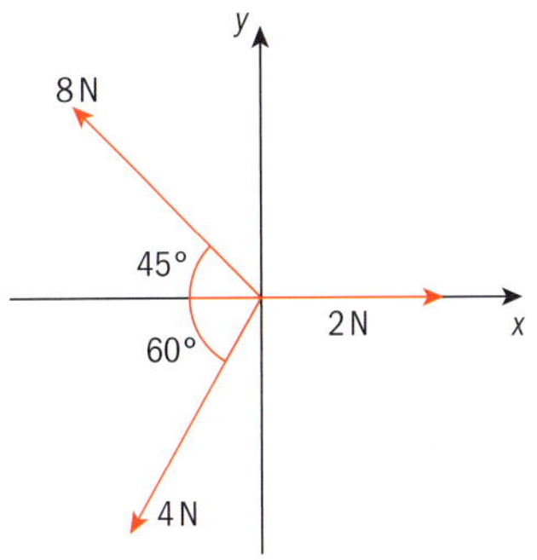

c)

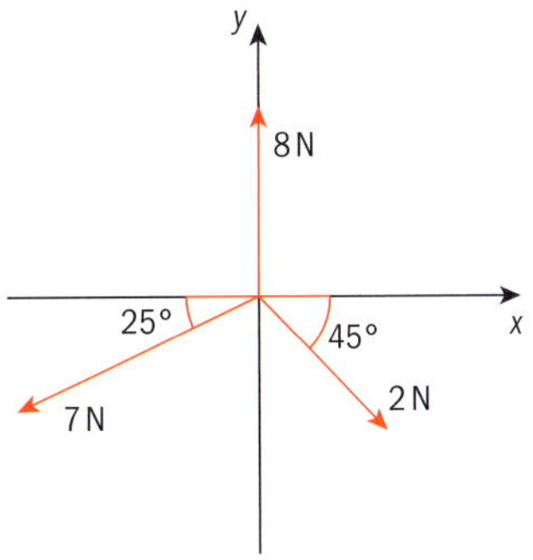

d)

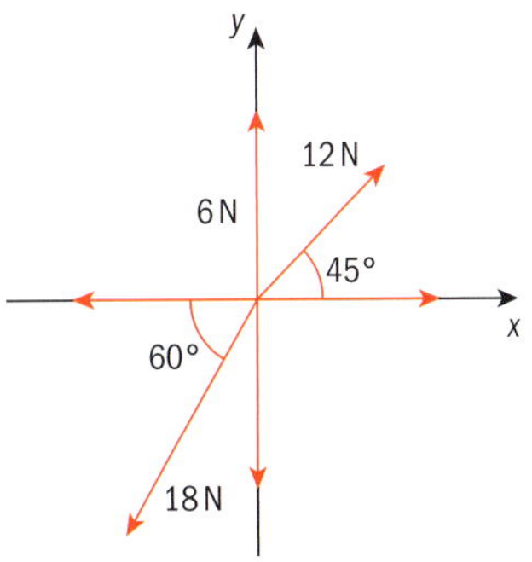

e)

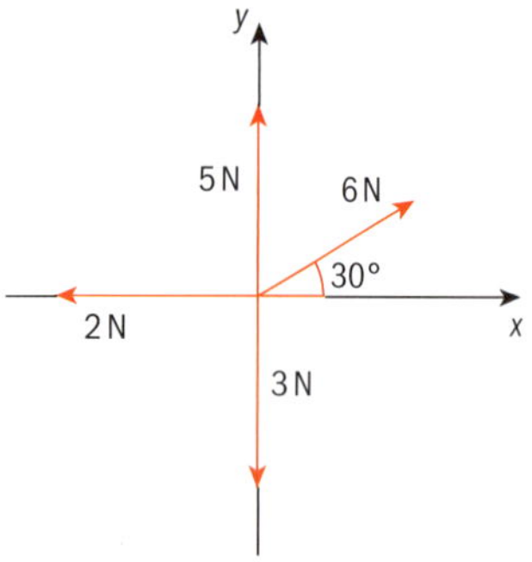

f)

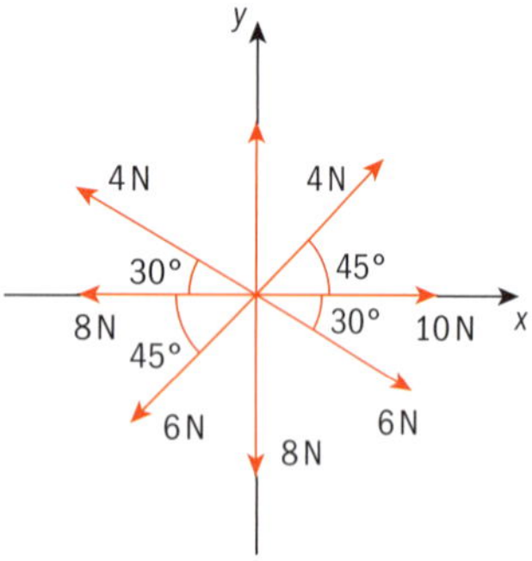

3. For the following find the resultant and the angle it makes with the positive x-axis.

a)

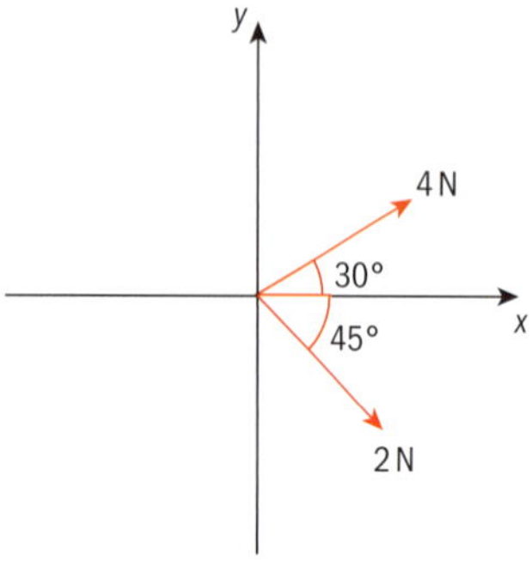

b)

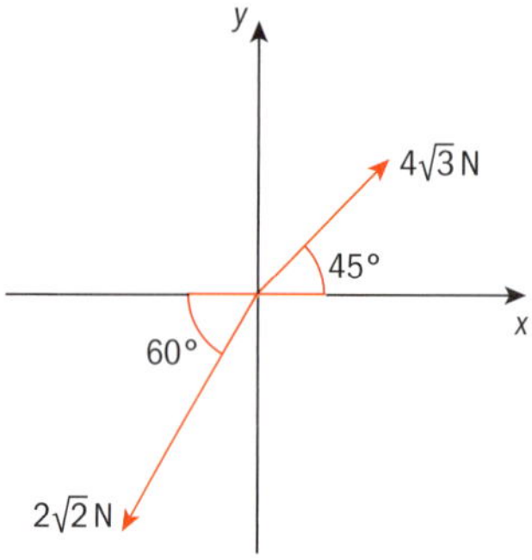

c)

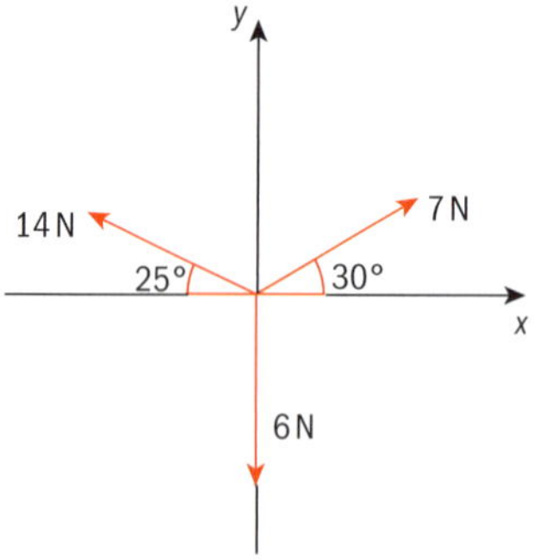

d)

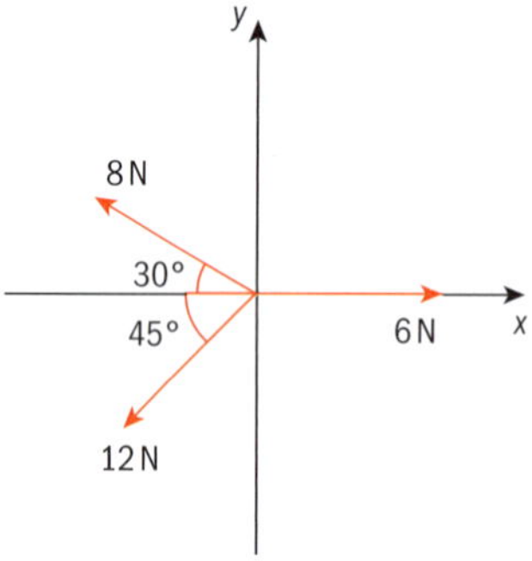

4. The angle between two forces of 350 N and 250 N is 35°. Find their resultant.

5. The angle between a force of 47 N and 56 N is 90°. Find their resultant.

6. The angle between a force of 200 N and XN is 20°. Given that their resultant has magnitude of 367.4 N, find the value of X.

7. Find the magnitude of the missing force, given that the resultant between the forces of 8 N and X N is $\sqrt{97}$ N, and the angle between the two forces is 60°.

8. Find the angle between a force of 10 N and 9 N, given that their resultant is 17.86 N.

9. The angle between a force of 5 N and X N is 90°. Given that their resultant has magnitude of $\sqrt{55}$ N, find the value of X.

EXAM-STYLE QUESTION

10.

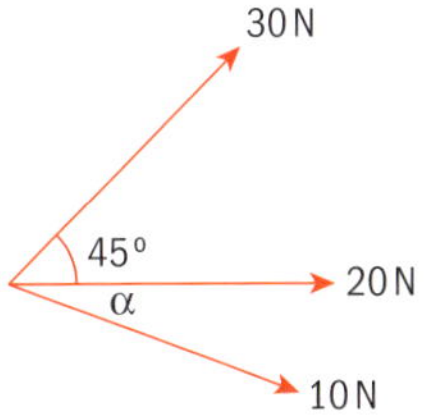

Coplanar forces of magnitudes 30 N, 20 N and 10 N act at a point in the directions shown in the diagram. Given that $\sin\alpha = \frac{5}{13}$, find the magnitude and the direction of the resultant of the three forces.

11.

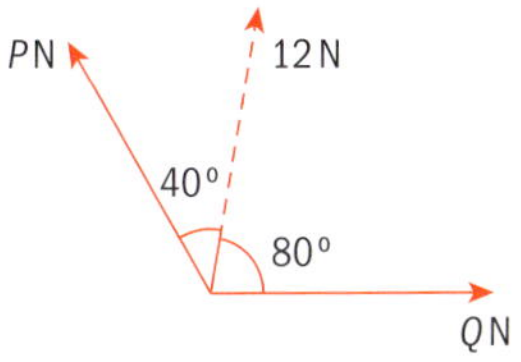

Two forces have magnitudes P N and Q N. The resultant of the two forces has magnitude 12 N and acts in a direction 40° clockwise from the force of magnitude P N and 80° anticlockwise from the force of magnitude Q N (see diagram). Find the value of Q.

Cambridge International AS and A Level Mathematics 9709, Paper 41 Q3 October/November 2009

12.

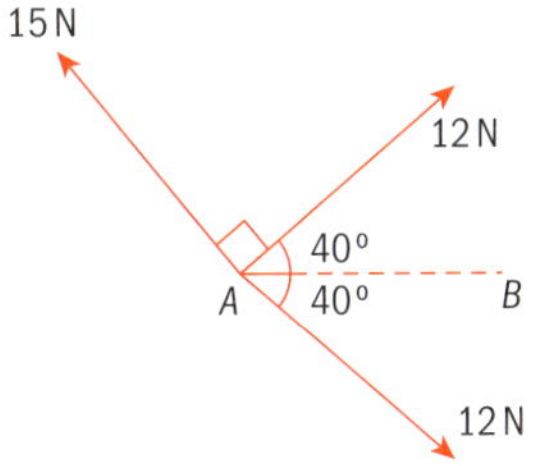

Three coplanar forces of magnitudes 15 N, 12 N and 12 N act at a point A in directions as shown in the diagram.

i) Find the component of the resultant of the three forces
   a) in the direction of AB,
   b) perpendicular to AB. [3]

ii) Hence find the magnitude and direction of the resultant of the three forces. [3]

Cambridge International AS and A Level Mathematics 9709, Paper 41 Q3 October/November 2011

EXAM-STYLE QUESTION

13.

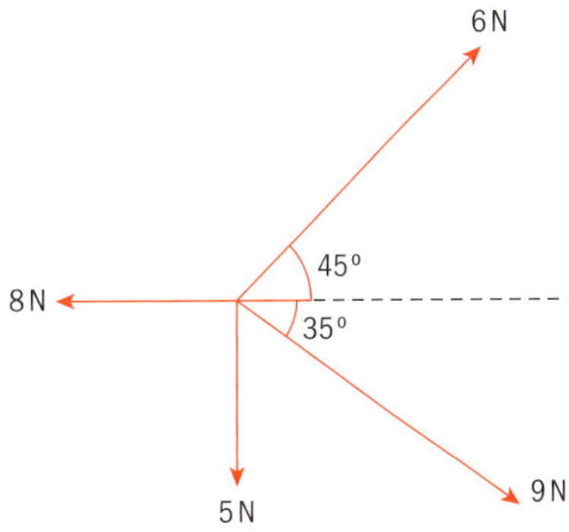

Four coplanar forces act at a point. The magnitudes of the forces are 8 N, 6 N, 9 N and 5 N, respectively, as shown in the diagram. Find the magnitude and direction of the result of the four forces.

14. 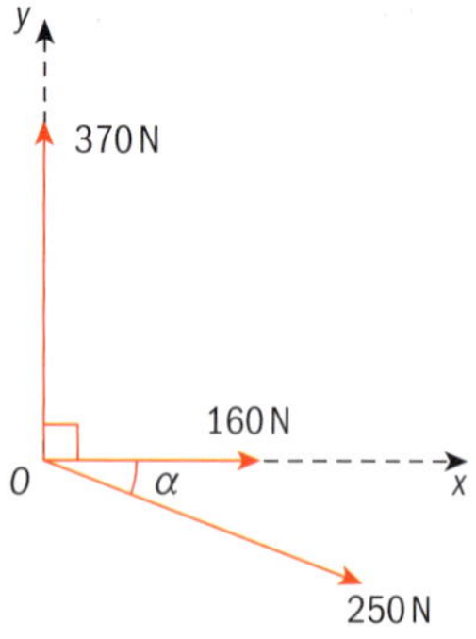

Coplanar forces of magnitudes 250 N, 160 N and 370 N act at a point O in the directions shown in the diagram, where the angle α is such that $\sin \alpha = 0.28$ and $\cos \alpha = 0.96$. Calculate the magnitude of the resultant of the three forces. Calculate also the angle that the resultant makes with the x-direction. [7]

Cambridge International AS and A Level Mathematics 9709, Paper 41 Q4 May/June 2010

15. Forces of magnitudes 13 N and 14 N act at a point O in the directions shown in the diagram. The resultant of these forces has magnitude 15 N. Find

(i) the value of θ, [3]

(ii) the component of the resultant in the direction of the force of magnitude 14 N. [2]

Cambridge International AS and A Level Mathematics 9709, Paper 41 Q2 May/June 2012

Chapter summary

- To find an angle between three known sides we use the cosine rule: $c^2 = a^2 + b^2 - 2ab \cos C$
- Component in the x-direction is $F\cos\theta$
- Component in the y-direction is $F\sin\theta$

Review exercise A

1. The piston on a steam engine moves forwards a distance of 2.5 metres and then returns. It takes 2 seconds for it to move forward and back at a constant speed.
   a) Sketch the displacement–time graph of the motion of the piston. (Consider the motion of a single point on the length of the piston.)
   b) Calculate the speed that the piston is moving.

2. A camera at the edge of an athletics track travels 100 metres in 10 seconds. Assume that it travels at a constant velocity for this time. At the end of the track the camera stops for 20 seconds before returning to its starting position in 20 seconds.
   a) Sketch the displacement–time graph of the camera as it moves up and down the track.
   b) Calculate the velocity of the camera as it moves in each direction.

3. A train moves out of a station travelling at $7\,\text{ms}^{-1}$. After 30 seconds it speeds up so that it is travelling at $18\,\text{ms}^{-1}$. It travels on for 7.5 minutes before slowing to $10\,\text{ms}^{-1}$. After a further 1.5 minutes it slows to $3\,\text{ms}^{-1}$ and finally comes to rest some 30 seconds later.
   a) Calculate the distance travelled for each stage of the train journey, the total distance and the average speed.
   b) Sketch a displacement–time graph of the train journey.

4. The displacement of a particle from O is S m at a time of $5T$ sec. The particle then returns to O arriving at time $7T$ sec.
   a) Sketch the displacement–time graph.
   b) If the initial velocity of the particle was $30\,\text{ms}^{-1}$, find the velocity of the particle on its return.
   c) The total time taken was 35 sec. Calculate the distance S.

5. A particle travels 15 m from a point O in a time T sec. It remains stationary for a further $2T$ sec before returning to O in T sec.
   a) Sketch the displacement–time graph.

   If the initial velocity of the particle was $20\,\text{ms}^{-1}$, calculate

   b) the value of T
   c) the total time taken to complete the journey.

6. Sapphira is running on the treadmill in the gym. Starting from rest she increases her speed to $6\,\text{ms}^{-1}$ in 30 sec. She runs at this speed for a further 2 min when she accelerates to $8\,\text{ms}^{-1}$ in 20 sec. Once she reaches that speed, she then slows to rest in a further 10 sec. Calculate
   a) the acceleration during the first 30 sec
   b) the acceleration from $6\,\text{ms}^{-1}$ to $8\,\text{ms}^{-1}$
   c) the deceleration at the end of her run
   d) the total distance that she runs on the treadmill.

7. A crane on a building site lifts material from the ground to the top of the building. The motion of the crane can be modelled in three phases. The first is acceleration from rest to $5\,\text{ms}^{-1}$. The second is motion with a uniform velocity and the third is deceleration to rest in 10 sec.

a) If the initial acceleration is $0.2\,ms^{-2}$, calculate the time that the crane is accelerating

b) If the material reaches the top of the building in 100 sec, calculate the deceleration and the height of the building.

8. A particle decelerates from a speed of $40\,ms^{-1}$ to a speed V in 10 sec. It travels at speed V for a further 10 sec and then decelerates to rest in 16 sec. The initial deceleration is $2\,ms^{-2}$.

a) Sketch a velocity-time graph for the particle.

b) Find the value of V.

c) Find the average velocity for the whole journey.

9. A particle decelerates from an initial velocity of $12\,ms^{-1}$ to a velocity V in T sec. It continues at speed V until 15 sec after it started. It then accelerates again so that 25 sec after the particle started its velocity is $20\,ms^{-1}$. The final acceleration is $1.8\,ms^{-2}$. The particle travels a total distance of 179 m.

a) Sketch a velocity-time graph for the particle.

b) Calculate the value of V.

c) Calculate the value of T.

10. A racing car is driven along a straight track that is 400 m long. It accelerates from a standing start and aims to complete the race in as short a time as possible. If the car passes the finishing line 6.5 sec after the start, find

a) the maximum velocity reached by the dragster

b) the acceleration of the car.

11.

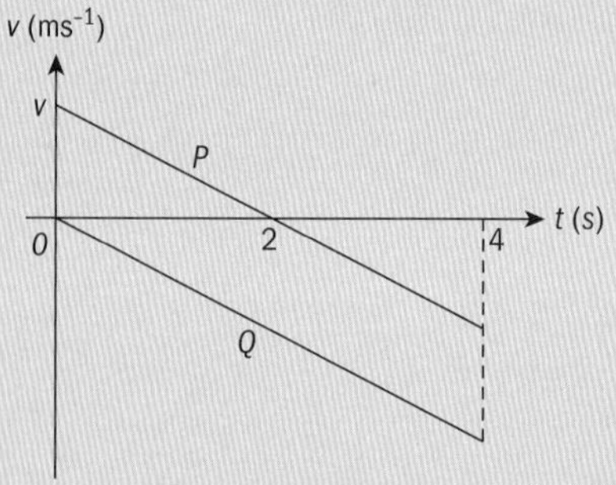

Two particles P and Q move vertically under gravity. The graphs show the upward velocity $v\,ms^{-1}$ of the particles at time t s, for $0 \le t \le 4$. P starts with velocity $V\,ms^{-1}$ and Q starts from rest.

i) Find the value of V. [2]

Given that Q reaches the horizontal ground when $t = 4$, find

ii) the speed with which Q reaches the ground, [1]

iii) the height of Q above the ground when $t = 0$. [2]

Cambridge International AS and A Level Mathematics 9709, Paper 41 Q1 October/November 2010

12. A train starts from rest at a station A and travels in a straight line to station B, where it comes to rest. The train moves with constant acceleration $0.025\,ms^{-2}$ for the first 600 s, with constant speed for the next 2600 s, and finally with constant deceleration $0.0375\,ms^{-2}$.

i) Find the total time taken for the train to travel from A to B. [4]

ii) Sketch the velocity-time graph for the journey and find the distance AB. [3]

iii) The speed of the train t seconds after leaving A is $7.5\,ms^{-1}$. State the possible values of t. [1]

Cambridge International AS and A Level Mathematics 9709, Paper 41 Q5 May/June 2011

13. A car accelerates uniformly in a straight line from rest to 30 ms^{-1} in 10 s. Calculate the acceleration of the car and the distance it travels in that time.

14. A, B and C are three points on a straight line in that order. The distance AB is 45 m, and BC is 32 m. A particle moves along the straight line with constant acceleration of 2 ms^{-2}. Given that it takes 5 *s* to travel from A to B, find

a) the speed of the particle at A

b) the time taken to travel from A to C.

15. A particle moves with constant acceleration of 0.4 ms^{-2} along a straight line passing through points A and B. It passes point B with a speed 0.8 ms^{-1} greater than its speed at A.

a) Given that the distance AB is 20 m, calculate the speed with which the particle passes the point A.

b) Find the time after passing A that the particle has a speed of 15 ms^{-1}.

16. A ball is thrown upwards with an initial speed of 2 ms^{-1} from a height of 1 m above horizontal ground. Find

a) the time when the speed of the ball is zero

b) the maximum height above the ground reached by the ball

c) the speed of the ball when it reaches the ground.

Assume that the acceleration due to gravity is 10 ms^{-2}.

17. A particle is projected vertically upwards from a fixed point O. The speed of projection is 5.6 ms^{-1}. At time T seconds after projection the particle is at a height of 0.92 m above O.

Find

a) the two possible values of T

b) the total time for which the particle is 0.92 m above O

c) the height of the particle, above O, when its speed is 2.8 ms^{-1}.

18. An object is released from rest at a height of 125 m above horizontal ground and falls freely under gravity, hitting a moving target P. The target P is moving on the ground in a straight line, with constant acceleration 0.8 ms^{-2}. At the instant the object is released, P passes through a point O with speed 5 ms^{-1}. Find the distance from O to the point where P is hit by the object.

Cambridge AS and A Level Mathematics 9709, Paper 41 Q1 October/November 2012

19. A particle slides down a smooth plane inclined at an angle of $\alpha°$ to the horizontal. The particle passes through the point A with speed 1.5 ms^{-1}, and 1.2 s later it passes through the point B with speed 4.5 ms^{-1}. Find

i) the acceleration of the particle,

ii) the value of α.

Cambridge AS and A Level Mathematics 9709, Paper 4 Q1 May/June 2008

20. A particle is projected vertically upwards from a point O with initial speed 12.5 ms^{-1}. At the same instant another particle is released from rest at a point 10 m vertically above O. Find the height above O at which the particles meet.

Cambridge AS and A Level Mathematics 9709, Paper 4 Q2 October/November 2007

21. The top of a cliff is 40 metres above the level of the sea. A man in a boat, close to the bottom of the cliff, is in difficulty and fires a distress signal vertically upwards from sea level. Find

i) the speed of projection of the signal given that it reaches a height of 5 m above the top of the cliff,

ii) the length of time for which the signal is above the level of the cliff.

The man fires another distress signal vertically upwards from sea level. This signal is above the level of the top of the cliff of the top of the cliff for $\sqrt{17}$ s.

iii) Find the speed of projection of the second signal.

Cambridge AS and A Level Mathematics 9709, Paper 41 Q3 May/June 2013

22.

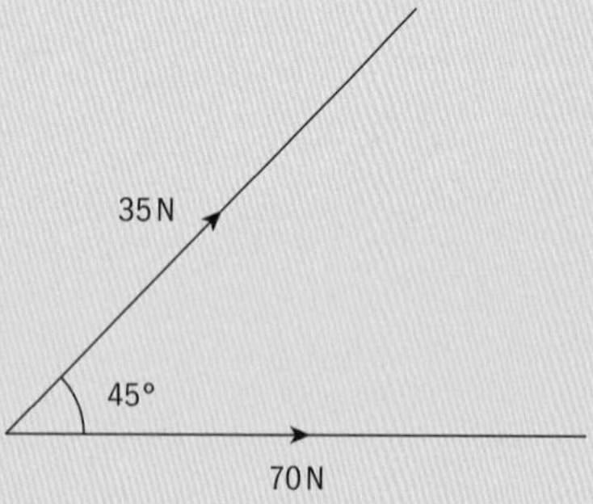

Find the resultant from the diagram.

23.

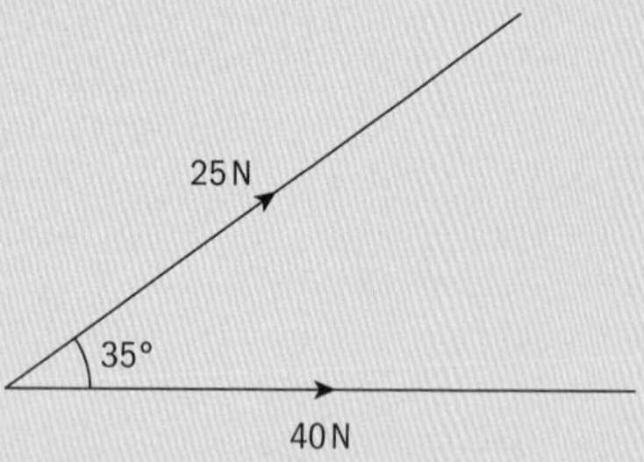

Find the resultant from the diagram.

24.

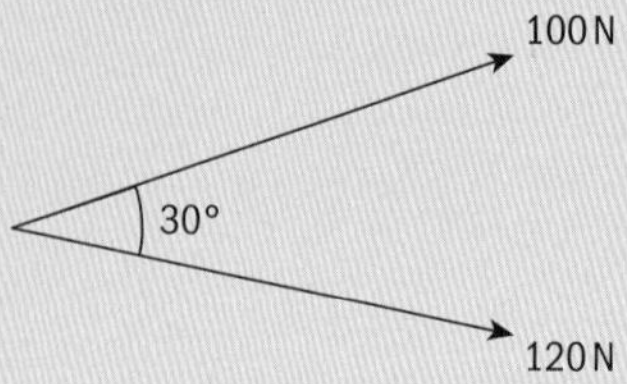

Find the resultant and the angle it makes with the larger of the two forces.

25.

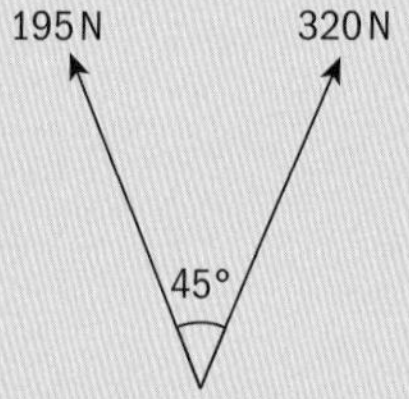

Find the resultant and the angle it makes with the larger of the two forces.

26.

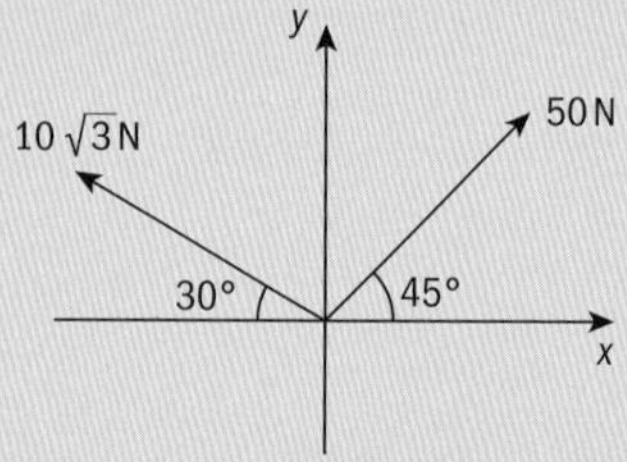

Find the components of the resultant of the given forces and express in the form **ai** + **bj**.

27.

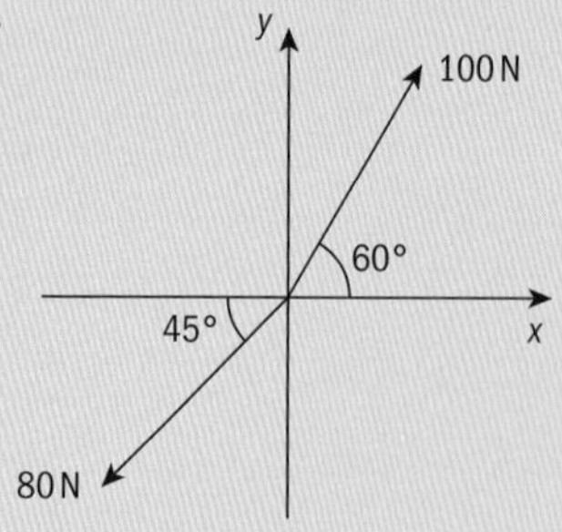

Find the components of the resultant of the given forces and express in the form **ai** + **bj**.

28.

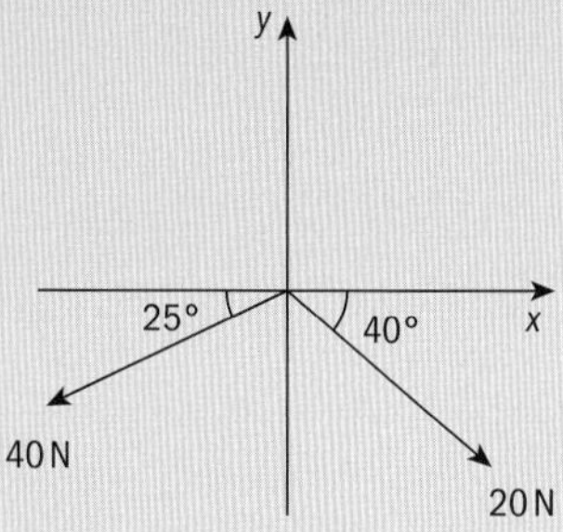

Find the components of the resultant of the given forces and express in the form **ai** + **bj**.

29.

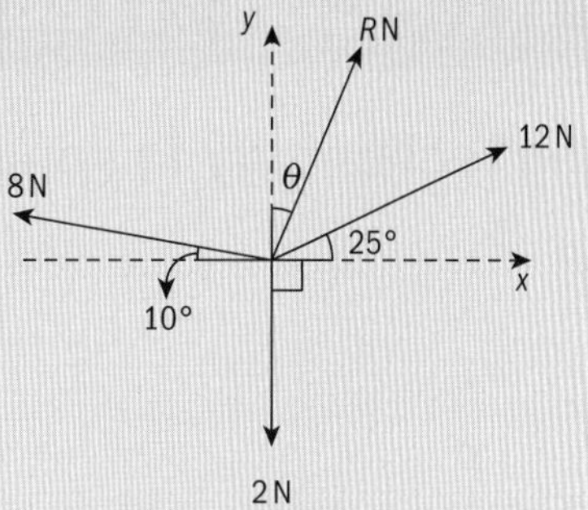

Three coplanar forces of magnitudes 8 N, 12 N and 2 N act at a point. The resultant of the forces has magnitude R N. The directions of the three forces and the resultant are shown in the diagram. Find R and θ. [7]

Cambridge International AS and A Level Mathematics 9709, Paper 41 Q4 October/November 2012

Maths in real-life

Challenging technology in sport

Tennis, cricket and soccer are all high-profile sports generating enormous amounts of money at the top level. All have turned to technology based on sophisticated application of some fairly simple mathematical ideas to help improve the decision making of their officials.

Tennis was the first to adopt the use of 3D computer simulation to predict the path of the ball based on information collected from multiple cameras. The US Open in 2006 was the first Grand Slam event to use it.

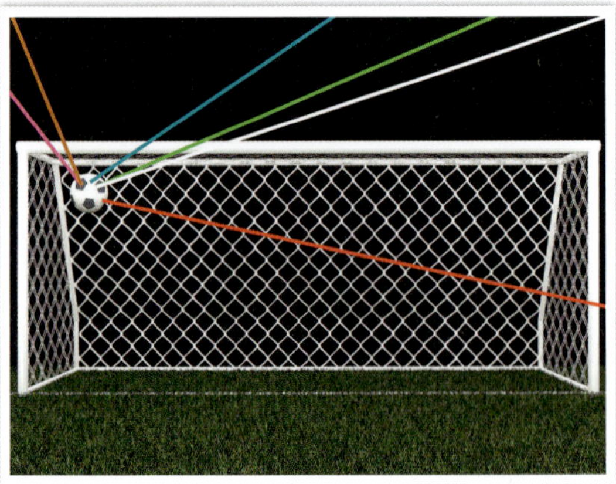

Soccer resisted the use of technology to aid officials' decision making until July 2012 when the World governing body approved its use in deciding whether a goal had been scored. The English Premier League implemented its use for the 2013–14 season and the World Cup in Brazil in 2014 was the first international tournament to use it.

How does it work?

Tennis normally uses 10 highspeed digital cameras which record the position of the ball repeatedly from different angles and heights. These are sent in real time to the central computer running the virtual reality software which uses the 2D images from different cameras to build up a 3D picture of the ball's position at each instant. The sequence of 3D positions is used to generate a trajectory for the ball which is used to predict its landing position.

▼ The requirements in cricket are much more complex. The first Test match to use it was between New Zealand and Pakistan in 2009.

Is it always used?

No – it is still extremely expensive to install and run it so it is only used at the very top level. The French Open is played on a clay surface which leaves a mark where the ball lands so it is not used there. The other tennis Grand Slams use the technology on one or more of the courts on which the most important matches are played.

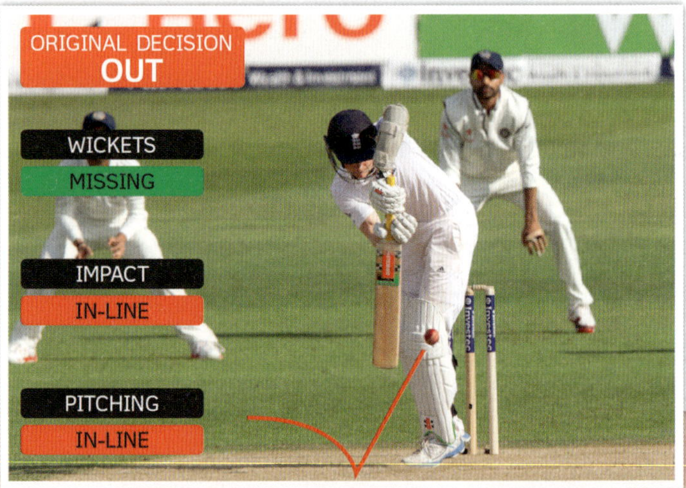

Is it reliable?

Manufacturers claim the average error in tennis is 3.6 mm (about the width of the fluff on the ball), and that the error is lower closer to the lines because that is where the technology is focused on. However, the computer prediction is relied on when a player makes a challenge – so if the computer says the ball is 1 mm out then it is called out even though that is much less than the reported margin of error.

Is it a good thing for sport?

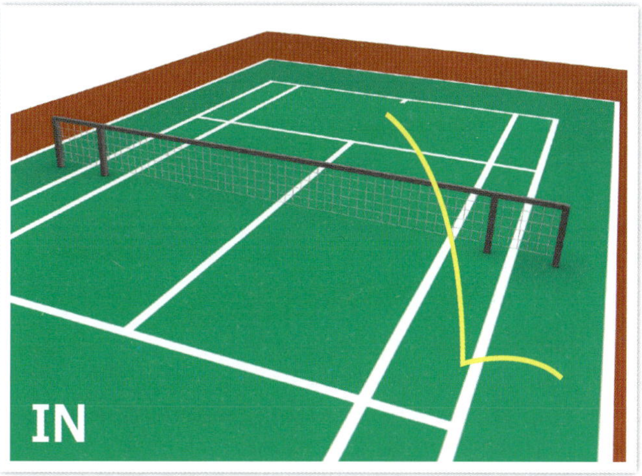

- It is not all good or all bad, so the answer to this is a matter of opinion. The accuracy of the technology is likely to improve with time because computers will be able to process data more quickly and the costs will reduce.
- Things to consider:
  - When is a decision challenged?
  - Are strategic challenges made?
  - What are the consequences of right / wrong decisions?
- The tennis scoring system is such that those 'big points', such as a game point or a match point, might have a huge influence on the outcome of the match!

4 Newton's laws

Soviet cosmonaut Yuri Gagarin became the first human to orbit the Earth. He achieved this in the Vostok spacecraft, which was launched on April 12, 1961. The spaceflight consisted of one orbit and took 108 minutes from launch to landing. At the end of the flight, Gagarin parachuted to the ground separately from his spacecraft after ejecting at an altitude of 7 km. Newton's laws explain how a spacecraft will accelerate back to Earth under the influence of gravity after being launched by a powerful force. We can also use Newton's laws to show the exact speed a spacecraft will need to achieve to go into orbit.

Objectives

- Use Newton's third law.
- Apply Newton's laws of motion to the linear motion of a particle of constant mass moving under the action of constant forces.
- Use the relationship between mass and weight.
- Solve simple problems which may be modelled as the motion of a particle moving vertically or on an inclined plane with constant acceleration.
- Solve simple problems which may be modelled as the motion of two particles, connected by a light inextensible string which may pass over a fixed smooth peg or light pulley.

Before you start

You should know how to:

1. Use the constant acceleration formulae.

$v = u + at$ $\qquad v^2 = u^2 + 2as$

$s = ut + \frac{1}{2}at^2$ $\qquad s = \frac{1}{2}(u + v)t$

e.g. Find s when $v = 15\,\text{ms}^{-1}$, $u = 3\,\text{ms}^{-1}$ and $a = 3\,\text{ms}^{-2}$.

$15^2 = 3^2 - 2 \times 3 \times s \quad s = 36\,\text{m}$

2. Resolve forces in perpendicular directions.

e.g. Find the horizontal and vertical components of a force of 15N at an angle of 30° to the horizontal.

Horizontal: $15\cos 30° = 13.0\,\text{N}$;

Vertical: $15\sin 30° = 7.5\,\text{N}$

Skills check:

1. a) Find s when $t = 2$, $u = 7$ and $a = 5$.

b) Find u when $v = 36$, $t = 3$ and $a = 8$.

c) Find s when $u = 5$, $v = 10$ and $a = 15$.

2. Find the horizontal and vertical components of a force of 12 N at 30° to the horizontal.

4.1 Newton's laws

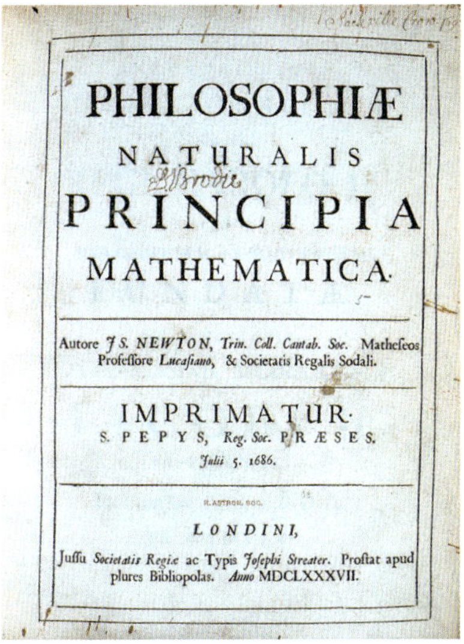
PHILOSOPHIÆ
NATURALIS
PRINCIPIA
MATHEMATICA
Autore J S. NEWTON, Trin. Coll. Cantab. Soc. Matheseos Professore Lucasiano, & Societatis Regalis Sodali.
IMPRIMATUR.
S. PEPYS, Reg. Soc. PRÆSES.
Julii 5. 1686.
LONDINI,
Jussu Societatis Regiæ ac Typis Josephi Streater. Prostat apud plures Bibliopolas. Anno MDCLXXXVII.

In ancient Greece the universe was understood in terms of the theories of Aristotle and Ptolemy. In their understanding, the Earth was at the centre of the universe. This view was held in the Western world until it was challenged by **Nicolaus Copernicus (1473–1543)** who first understood how the Earth revolved around the Sun, **Johannes Kepler (1571–1630)** who studied planetary motion, and **René Descartes (1596–1650)** who applied mathematical principles to mechanical theory.

Finally, classical mechanics, the study of the motion of bodies, was founded by **Sir Isaac Newton's** publication of *Philosophiæ Naturalis Principia Mathematica* (1686). In this book, Newton first stated his three laws of motion. Newton's laws have been used now for over three centuries. These laws have been restated and added to, and new theories such as those of quantum physics have emerged. Newton's laws, however, are still the cornerstones of mechanics.

Newton's first law

Every body remains in a state of rest or of uniform motion in a straight line unless external forces act on it.

So if a car is parked in the street, it will not move unless it is acted on by the thrust of the engine, or it is pushed or blown along by a hurricane for example. So why does the driver need to keep their foot down on the accelerator pedal in order to maintain uniform motion? It is because the force is the resultant of all the forces acting on the car. There is the resistance of the road, the force due to gravity if the car is travelling up a hill, and so on.

Newton's second law

The resultant of the forces acting on a body is equal to the mass of the body multiplied by its acceleration in the direction of that force.

This law can be stated as $F = ma$. More properly, both F and a should be **vectors** (i.e. they have both magnitude and direction).

A **newton** is defined as the force which is needed to give a mass of 1 kg an acceleration of $1\ \text{ms}^{-2}$.

The **weight** of an object is the force that gravity exerts on it. A free falling object, that is an object falling solely under the influence of gravity, has an acceleration of approximately $10\ \text{ms}^{-2}$ downwards towards the earth.

This value, the acceleration due to gravity, is known as g. There are slight variations of this value, due mainly to altitude. The value would also be different, for example, on the moon.

The weight of an object of mass m is then mg according to Newton's second law. The units of weight are the same as the units of a force – newtons. Hence a body of mass 20 kg will have a weight of 200 N.

Newton's third law

To every action there is an equal and opposite reaction.

A car parked on the street has a weight that is acting downwards on the surface of the street. The car is prevented from disappearing through the surface of the road by the reaction of the street, which acts in opposition to the car's weight and keeps the car in equilibrium. The weight of the car, mg, and the normal reaction, R, are equal and opposite forces.

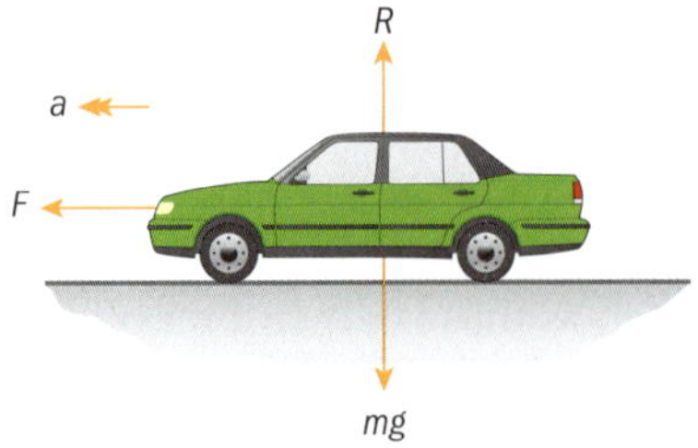

Note: In mechanics we talk about **bodies**, for example a cricket ball, a car, or a spacecraft. Strictly, Newton's three laws apply to **particles**, however we use them for most cases involving bodies because they work well enough in most situations and are easy to calculate.

Did you know?

According to legend, Newton discovered the universal law of gravitation when he was hit on the head by an apple while sitting under a tree. Like all legends, it may not be entirely true, but Newton did come up with some important ideas about gravity. The most important for us is that the **acceleration due to gravity** of all objects near the earth's surface is a constant, g, which is approximately 9.81 ms^{-2}, and we often round this value up to 10 ms^{-2} in calculations.

The Italian scientist and philosopher **Galileo Galilei (1564–1642)** had shown by dropping two balls of different masses from the Leaning Tower of Pisa that the times they took to reach the ground (and thus their acceleration) were the same. Therefore the acceleration is independent of the mass of the body falling. However, the value of g would be different on the Moon or on a spacecraft orbiting the Earth.

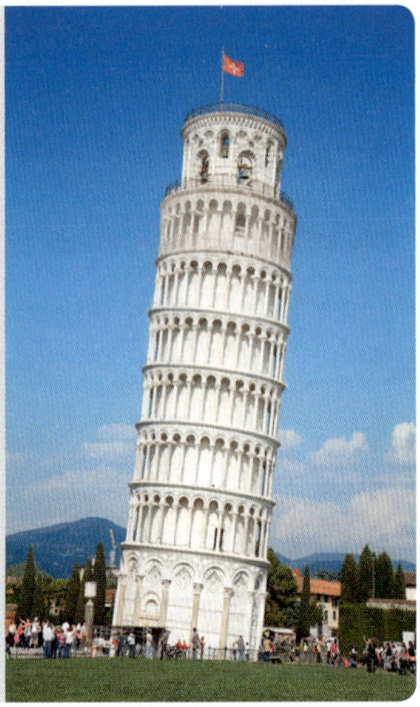

Example 1

A particle of mass m kg hangs on a light, inextensible string. The acceleration due to gravity is g. Calculate

a) the weight W of the particle

b) the tension T in the string.

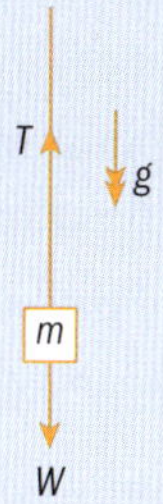

a) Applying Newton's second law vertically,

$W = mg$

b) Applying Newton's third law vertically,

$T = W$

$T = mg$

Example 1 demonstrates a very important result to do with the weight of a particle.

The mass m of a particle is measured in kg. The weight of that body is a force, mg, and is measured in newtons. As stated previously, 1N is the force that would give a mass of 1 kg an acceleration of 1 ms^{-2}. The weight of a particle could change if it moved somewhere where the acceleration due to gravity was different. Its mass, however, does not change. An astronaut in space could become **weightless** but he or she would still have the same mass as they do on Earth.

In Example 2 and Exercise 4.1, we look at situations where forces act either horizontally or vertically.

Example 2

A particle of mass 3 kg rests on a smooth plane. It is pulled by a horizontal force of 4 N.

Taking the value of g as 10 ms^{-2}, calculate

a) the horizontal acceleration of the particle, a

b) the normal reaction, R.

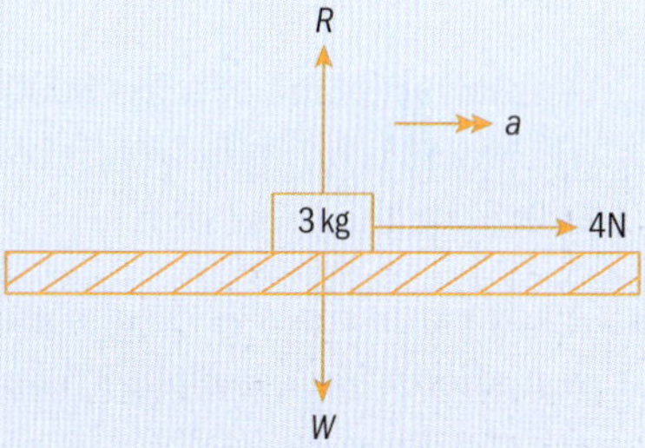

a) Applying Newton's second law horizontally

$4 = 3a$

$a = 1.33\ ms^{-2}$

b) Applying Newton's second law vertically,

$W = 3g$

$\quad = 3 \times 10$

$W = 30\ N$

Applying Newton's third law vertically,

$R = W$

$R = 30\ N$

Exercise 4.1

1. An ice cube rests on a horizontal table in the carriage of a train. Explain what happens to the ice cube when the train is
   a) accelerating out of the station
   b) travelling at constant velocity
   c) decelerating on approach to the next station.
2. A sledge stands on smooth horizontal icy ground. The sledge has a mass of 35 kg. If a force of 105 N acts on the sledge, find its acceleration.
3. A man pushes a box on horizontal ground with a force of 240 N. There is a frictional force of 60 N opposing the motion. If the acceleration of the box is 12 ms^{-2}, calculate the mass of the box in kg.

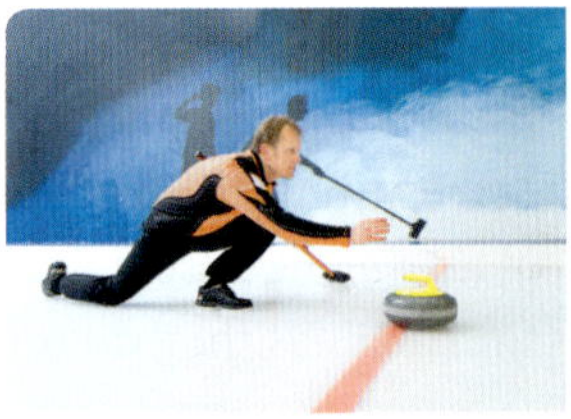

4. In the game of curling, a heavy granite stone of mass 20 kg is slid across an ice rink. Ignoring friction, if the acceleration of the stone is 2.4 ms^{-2}, calculate the force with which the stone is thrown.
5. A wooden block of mass 5 kg is at rest on a smooth horizontal table, 1.6 m from the edge of the table. The block is pulled towards the edge by a horizontal string. The tension in the string is 1 N. Calculate the time taken to reach the edge of the table.
6. A small seaplane of mass 8000 kg is travelling at v ms^{-1} and lands on the sea. The plane is brought to rest by water resistance of 960 N in 600 m. Calculate the value of v.
7. A toy car of mass 0.35 kg is moving at a velocity of 2 ms^{-1} and comes to rest after it has travelled 3.5 m. Calculate the resistance force that slows the car down.
8. A porter at a railway station is dragging a suitcase of mass 82.5 kg along the platform with an acceleration of 0.175 ms^{-2}. The horizontal force that he exerts is 170 N. Find the frictional force between the floor and the trunk.
9. A particle of mass 2.5 kg is pulled along a horizontal surface by a string parallel to the surface with an acceleration of 2.8 ms^{-2}. Given that there is a frictional force of 4N that opposes the motion of the particle, find the tension in the string. When the particle is travelling at a speed of 3 ms^{-1}, the string breaks. Calculate how much further the particle will travel before coming to rest.

10. A goods lift has a mass of 750 kg and can hold a maximum load of 1200 kg. The lift is raised and lowered by a cable. Using $g = 10\,\text{ms}^{-2}$, calculate

a) the tension in the cable when it is being raised with a full load and an acceleration of $0.5\,\text{ms}^{-2}$

b) the tension in the cable when the empty lift is being lowered with an acceleration of $0.7\,\text{ms}^{-2}$

c) the mass of the load if there is a tension of 15 200 N when the partially loaded lift is being raised at constant speed.

4.2 Resolving when on an inclined plane

In section 3.3, we resolved a single force into perpendicular components. We are now going to take this idea and apply it to the weight of a body that is resting on an inclined plane.

Inclined plane means 'slope'.

A body of mass m rests on a plane inclined at an angle θ to the horizontal. The weight of the body is mg acting downwards where g is the acceleration due to gravity.

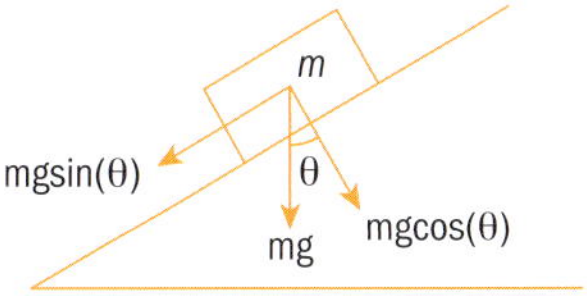

We are going to resolve the weight of the body into two components parallel to and perpendicular to the plane.

Look carefully at the two equal angles in the triangle below the force diagram. The angle of inclination of the plane is equal to the angle between the downward vertical and the perpendicular to the plane.

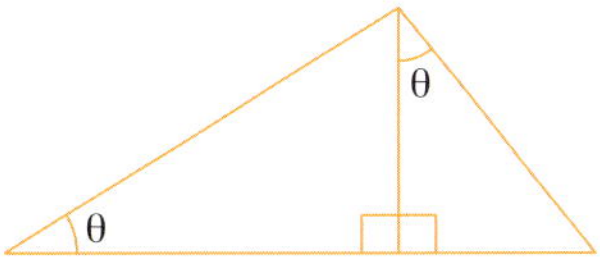

The two components of the weight are therefore:

$mg\cos(\theta)$ perpendicular to the plane, and

$mg\sin(\theta)$ parallel to the plane

Example 3

A body of mass 10 kg rests on a plane that is inclined at an angle of 30° to the horizontal.

Find the components of the weight of the body

a) parallel to the plane

b) perpendicular to the plane.

(Take g as $10\,\text{ms}^{-2}$)

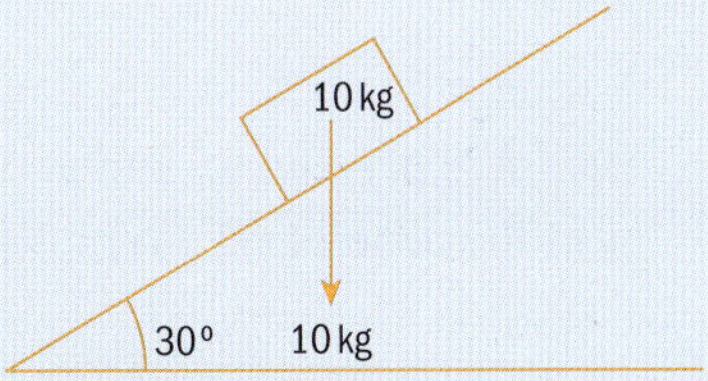

Continued on the next page

The weight of the body is $10 \times 10 = 100\,\text{N}$

a) Component parallel to the plane

$100 \sin 30° = 50\,\text{N}$

b) Component perpendicular to the plane

$100 \cos 30° = 86.6\,\text{N}$ (3 s.f.)

In the following exercise take g as $10\,\text{ms}^{-2}$.

Exercise 4.2

1. A truck carrying fruit for the local market breaks down on a hill. Given that the truck and fruit have mass 1500 kg, calculate the forces both parallel and perpendicular to the surface of the hill

i) down the plane

ii) at right angles to the plane, if the hill is at

a) 5° **b)** 35° **c)** 78°

2. A bookseller is delivering some heavy books. While out on her journey, she stops to talk to a customer. As she does this, the box containing the books, which has mass 200 kg, is at rest on a hill. Given that the force acting on the box parallel to the hill is 407.51 N, find the angle of incline.

3. A car with three passengers on board is being driven up a hill when it suddenly breaks down. The hill is inclined at 40°. If the total mass of the car and passengers is 2200 kg, calculate the force parallel to the hill and the force perpendicular to the hill.

One by one, each passenger gets out of the car and does not come back. The mass of each passenger in the order they leave the car is 220 kg, 190 kg and 150 kg, respectively.

a) Find the component of weight both down the hill and perpendicular to the hill before and after *each* person has left the car.

b) Find the difference between the initial force perpendicular to the hill and the final force perpendicular to the hill.

4. A skier falls while skiing down a slope, and is left lying on the ground. The force parallel to, and acting down the slope is 780.81 N at an angle of indirection of 60°. What is the mass of the skier?

4.3 Multiple forces

In 4.1, we looked at the motion of a particle or a body when acted on by forces in one direction only. When the forces act in different directions, it is necessary to resolve the forces to be able to calculate their effect. In some cases we resolve them horizontally and vertically, and in other cases we might look at forces acting up and perpendicular to a plane.

Example 4

A particle of mass 25 kg rests on a smooth inclined plane which is at 31° to the horizontal. It is pulled by a force of 300 N up the plane.

Taking the value of g as $10\,\text{ms}^{-2}$, calculate

a) the acceleration of the particle up the plane, a

b) the normal reaction, R.

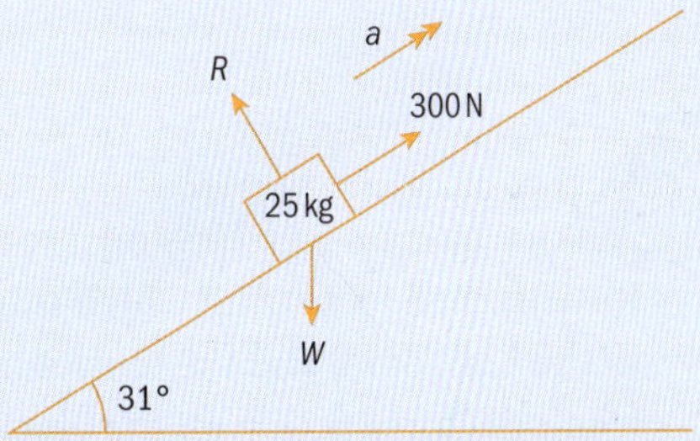

a) $W = mg$

$W = 25 \times 10$

$W = 250\,\text{N}$

Resolving forces up the plane, using Newton's second law,

$300 - 250 \times \sin 31° = 25a$

$171.2 = 25a$

$a = 6.85\,\text{ms}^{-2}$

b) Resolving forces perpendicular to the plane, using Newton's third law,

$R = 250 \cos 31°$

$R = 214\,\text{N}$

Exercise 4.3

1. The diagram shows the forces, all in newtons, which act on an object which is at rest. Find the values of the forces marked P and Q.

a)

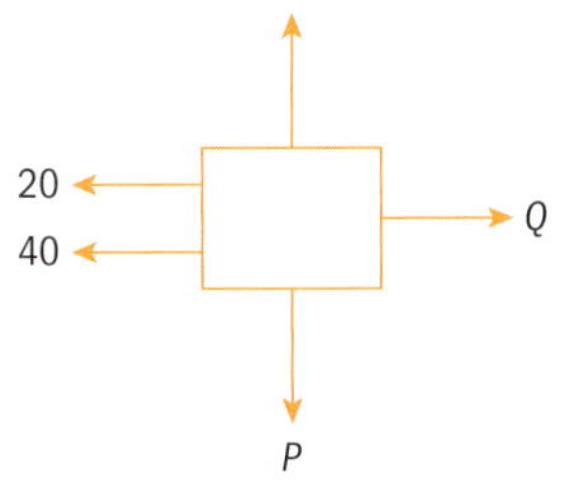

b)

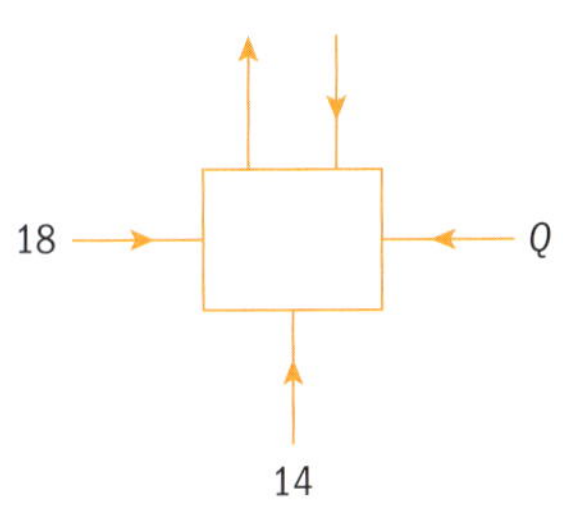

2. The following forces, in newtons, act on a body: $-3\mathbf{i} + 4\mathbf{j}$, $6\mathbf{i} - 2\mathbf{j}$, $4\mathbf{i} + 5\mathbf{j}$ and $-7\mathbf{j}$. Determine whether the body is accelerating and, if so, in which direction.

3. A box is being pulled up a smooth slope with an acceleration a by a light inextensible string. Show, on a copy of the following diagram, all the forces that are acting on the box.

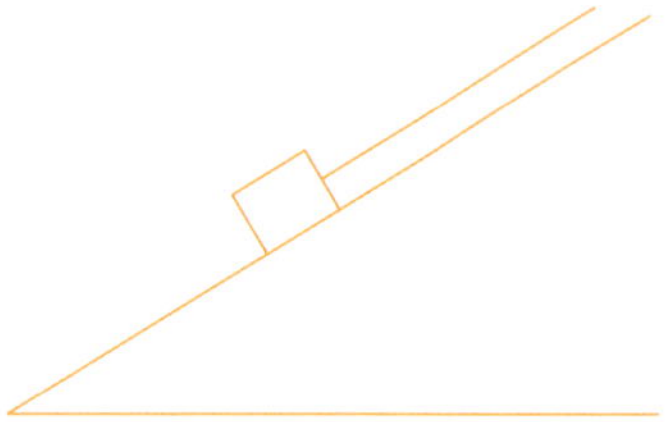

4. A block of mass 5 kg rests on a rough horizontal plane. It is attached to a string which is inclined at an angle of 30° to the horizontal and the tension in the string is 27 N. Find the frictional force acting on the block.

5. A small block of mass 3 kg is being pulled by a string up a rough plane which exerts a frictional force of 2.5 N and which is inclined at 30° to the horizontal. If the block accelerates at $0.5\ \mathrm{ms^{-1}}$, calculate, in terms of g, the tension in the string.

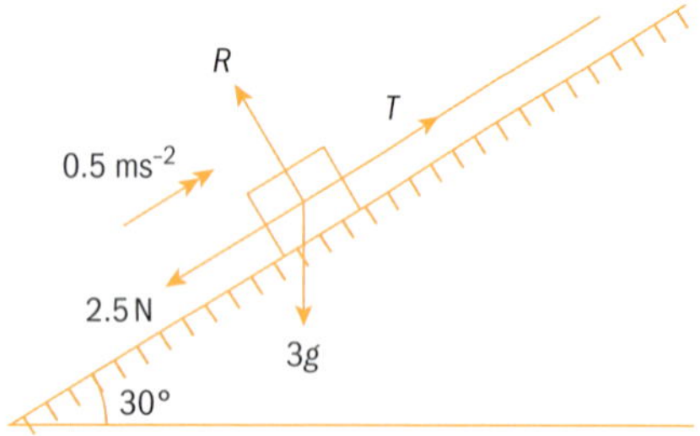

6. A boat is being pulled along a canal by two cables attached to two horses. The mass of the boat is 800 kg. One horse pulls with a force of 35 N and the cable is at an angle of 10° to the banks of the canal. The other pulls with a force of 40 N. Find the angle the second horse should pull at in order that the boat moves forward and does not get dragged sideways. Ignoring any frictional resistance, find the acceleration of the boat.

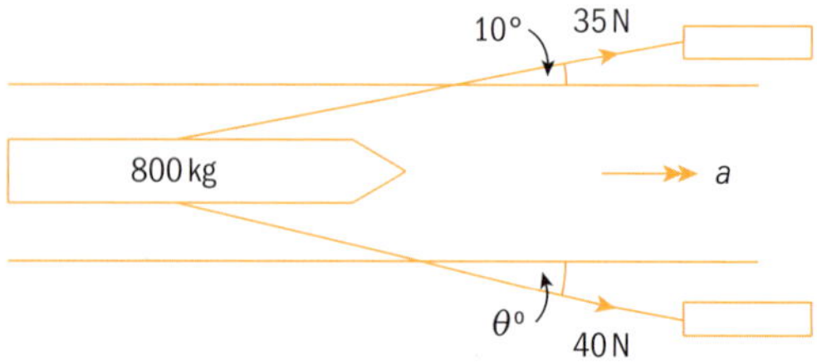

7. A body of mass 5 kg is sliding down a smooth plane inclined at an angle of 25° to the horizontal. Its speed is being controlled by a rope that is inclined to the plane at an angle of 10° to the plane. The tension in the rope is 8 N. The body is released from rest and slides down the plane. Taking g as $10\,\text{ms}^{-2}$, calculate how far it will travel in 6s.

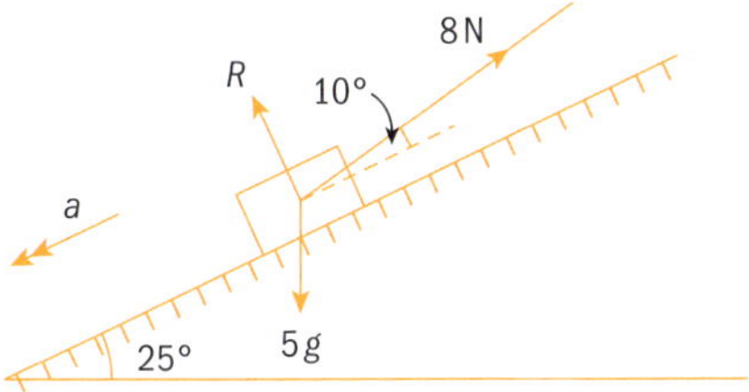

8. A paraglider of mass 85 kg is pulled by a rope attached to a speedboat. With the rope making an angle of 20° to the horizontal the paraglider moves in a straight line parallel to the water surface with an acceleration of $1.5\,\text{ms}^{-2}$. The tension in the rope is 240 N. Taking g as $10\,\text{ms}^{-2}$, calculate the magnitude of the vertical lift force acting on the glider and the magnitude of the air resistance.

4.4 Connected particles

In this third section we look at particles connected by a light inextensible string that passes over a smooth pulley.

- A **light inextensible string** is a string whose length remains the same whether motion is taking place or not.
- A **pulley** is a wheel over which a string passes. A **smooth pulley** has no friction in its bearings.

Note: Friction will be discussed in Chapter 6.

In real-life, strings are not always light and inextensible, and not all pulleys are smooth, but we make assumptions when modelling to make calculations more manageable. In these examples, the forces acting on the particles are their weight and the tension in the string. When a light inextensible string passes over a smooth, light pulley or over a smooth peg, the tension in the string either side of the pulley will be the same, i.e. the tension will be constant along its length. We also find that, because the particles are connected, that their acceleration will have the same magnitude, but in opposite directions.

Example 5

Two particles of mass 5 kg and 7 kg are connected by a light, inextensible string, which hangs over a smooth, light pulley.

Taking the value of g as $10\,\text{ms}^{-2}$, calculate

a) the weights of the two particles, W_1 and W_2

b) the acceleration of the system, a

c) the normal reaction at the point where the pulley is suspended, R.

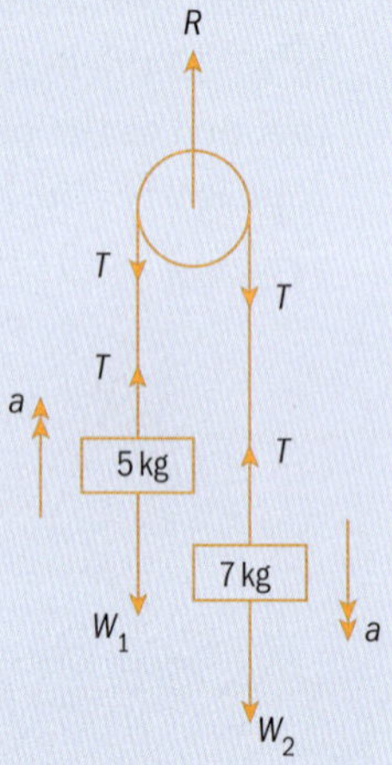

a) $W_1 = 5g$

$W_1 = 5 \times 10 = 50\,\text{N}$

$W_2 = 7g$

$W_2 = 7 \times 10 = 70\,\text{N}$

b) Resolving forces vertically for each of the particles,

$T - 50 = 5a$

$70 - T = 7a$

$20 = 12a$

$a = 1.67\,\text{ms}^{-2}$

c) $R = 2T$

$R = 2 \times (50 + 5 \times 1.67)$

$R = 117\,\text{N}$

Exercise 4.4

1. Some repairs are being carried out on a tall building. A pulley is attached to the top of the scaffolding and a rope runs over the pulley with buckets of mass 2 kg attached to each end. One bucket, at the top of the building, is filled with rubble of mass 8 kg and released. What will be its acceleration? Give your answer in terms of g.

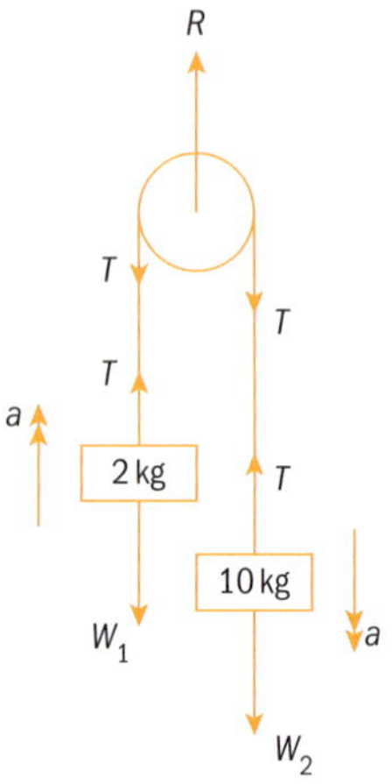

2. A small block of mass 4 kg rests on a table and is connected by a light inextensible string that passes over a smooth pulley fixed to the edge of the table, to another small block of mass 2 kg which is hanging freely. If the table is rough and exerts a frictional force of $1.5gN$ on the block, find, in terms of g, the acceleration of the system and the tension in the string.

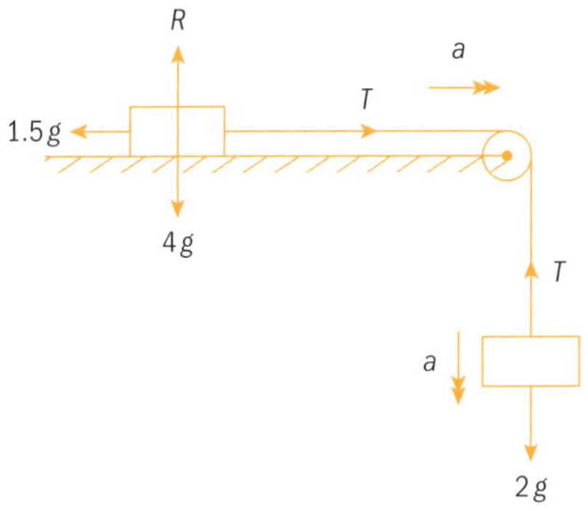

3. A block of mass 15 kg rests on a plane that is inclined at 30° to the horizontal. It is connected by a light inextensible string, that passes over a smooth pulley that is attached to the top of the plane to another block of mass 10 kg which is hanging freely. If the frictional force exerted by the plane on the block is 2.4 N, find the resulting acceleration of the system. (Take g as $10\,\text{ms}^{-2}$.)

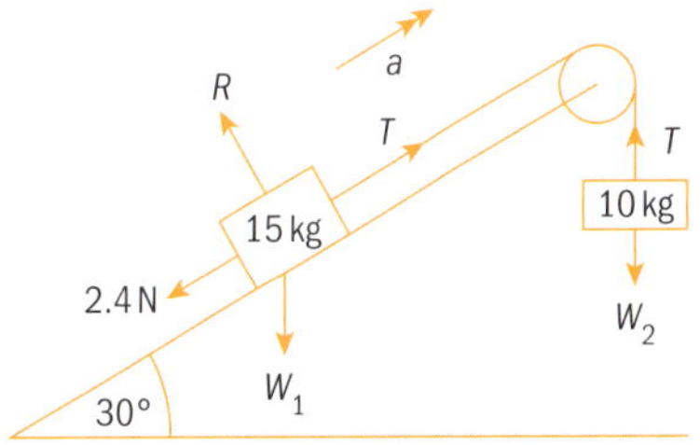

4. A funicular railway runs up a steep slope and has two carriages connected by a cable that runs over a pulley at the top of the slope. The weight of one carriage running down the track helps to pull up the second carriage. Each carriage weighs 10 000 kg and the track is inclined at an angle of 40° to the horizontal. The frictional resistance of the track is 100 N. Ignoring the weight of the cable and resistance in the pulley, calculate the force required to pull a carriage up the slope with an acceleration of $2.0\,\text{ms}^{-2}$ (labelled F in the diagram below).

Did you know?
A funicular railway is a type of railway where a cable is used to pull two trams up or down an inclined plane. As one tram ascends, the other descends and they counterbalance each other.

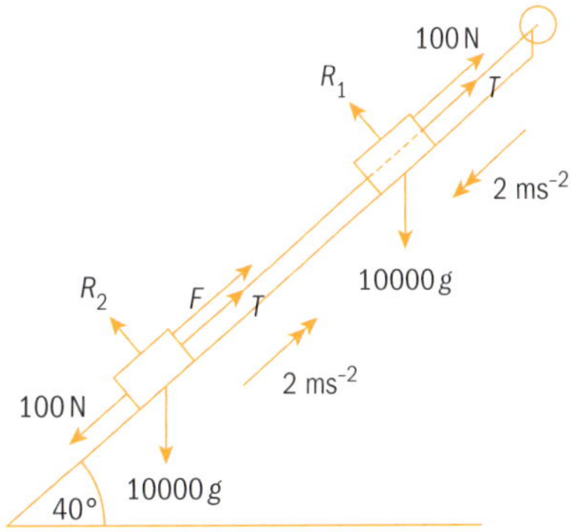

Summary exercise 4

Use $g = 10\,\text{ms}^{-2}$ where a numerical value is required.

1. Particles A of mass 0.65 kg and B of mass 0.35 kg are attached to the ends of a light inextensible string which passes over a smooth pulley. B is held at rest with the string taut and both of its straight parts vertical. The system is released from rest and the particles move vertically. Find the tension in the string and the magnitude of the resultant force exerted on the pulley by the string. [5]

Cambridge International AS and A Level Mathematics 9709, Paper 41 Q2 October/November 2011

2. A lift of mass 750 kg is suspended by a cable. A passenger of mass 65 kg stands in the lift.
   a) When the lift is accelerating upwards at $0.4\,\text{ms}^{-2}$, find
      i) the force exerted by the passenger on the floor of the lift
      ii) the tension in the cable.
   b) When the lift is accelerating downwards at $0.6\,\text{ms}^{-2}$, find
      i) the force exerted by the passenger on the floor of the lift
      ii) the tension in the cable.

3. Particles P and Q, of masses 0.4 kg and 0.25 kg respectively, are attached to the ends of a light inextensible string. P is held at rest on a rough horizontal table with the string passing over a smooth pulley at the edge of the table. Q hangs vertically below the pulley. The system is released and Q starts to move downwards with acceleration $2\,\text{ms}^{-2}$.
   Find
   a) the tension in the string after the system is released
   b) the frictional force acting on P.

4. A car of mass 1200 kg moves in a straight line along horizontal ground. The resistance to motion of the car is constant and has magnitude 960 N.
   a) Calculate the acceleration of the car when the car's engine is providing a driving force of 1440 N.

   When the car passes through a point it is moving with a constant speed of $18\,\text{ms}^{-1}$. At that instant the engine is switched off. The car continues along the straight line until it comes to rest.

   b) Find the distance it takes to come to rest.

5.

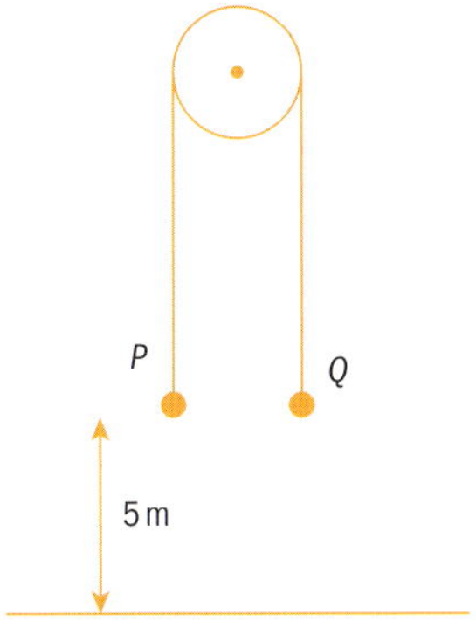

Particles P and Q, of masses 0.55 kg and 0.45 kg respectively, are attached to the ends of a light inextensible string which passes over a smooth fixed pulley. The particles are held at rest with the string taut and its straight parts vertical. Both particles are at a height of 5 m above the ground (see diagram). The system is released.

i) Find the acceleration with which P starts to move. [3]

The string breaks after 2 s and in the subsequent motion P and Q move vertically under gravity.

ii) At the instant that the string breaks, find

a) the height above the ground of P and of Q, [2]

b) the speed of the particles. [1]

iii) Show that Q reaches the ground 0.8 s later than P. [4]

Cambridge International AS and A Level Mathematics 9709, Paper 41 Q6 October/November 2009

Chapter summary

Newton's first law

Every body remains in a state of rest or of uniform motion in a straight line unless external forces act on it.

Newton's second law

The resultant of the forces acting on a body is equal to the mass of the body multiplied by its acceleration in the direction of that force.

Newton's third law

To every action there is an equal and opposite reaction.

- The mass m of a particle is measured in kg. The weight of that body is a force, mg, and is measured in newtons.
- 1 N is the force required to give a mass of 1 kg an acceleration of 1 ms^{-2}.

Resolving when on an inclined plane

- On an inclined plane and perpendicular to the incline, $F = mg\cos\theta$
- On an inclined plane and parallel to the incline, $F = mg\sin\theta$

5 Equilibrium

Equilibrium is the state where two or more forces act upon a particle, or rigid body, and motion does not take place because the forces cancel each other out. The particle or rigid body is at rest. In engineering, equilibrium is extremely important for the stability of objects. From designing cars to constructing bridges, it is crucial to understand how forces balance.

Objectives

- Use the principle that, when a particle is in equilibrium, the vector sum of the forces acting is zero, or equivalently, that the sum of the components in any direction is zero.
- Solve simple problems which may be modelled as the motion of two particles, connected by a light inextensible string which may pass over a fixed smooth peg or light pulley.

Before you start

You should know how to:

1. Use Pythagoras' theorem to find the lengths of the sides of a right-angled triangle, e.g.

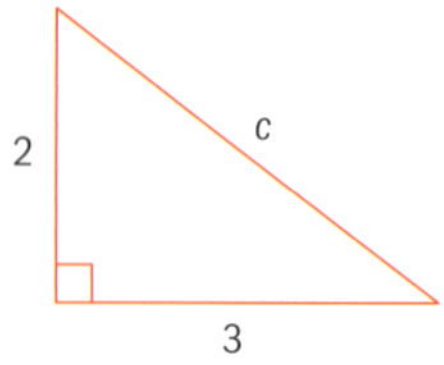

$c^2 = 2^2 + 3^2$

$\Rightarrow c = 3.61$ (2 d.p.)

2. Use the sine rule which relates the angles and sides of a triangle, e.g.

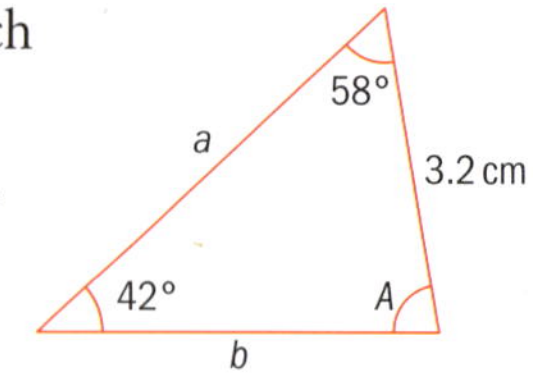

$$\frac{a}{\sin A} = \frac{b}{\sin 58} = \frac{3.2}{\sin 42}$$

$b = 4.06$ cm

3. Use the cosine rule which relates the angles and sides of a triangle, e.g.

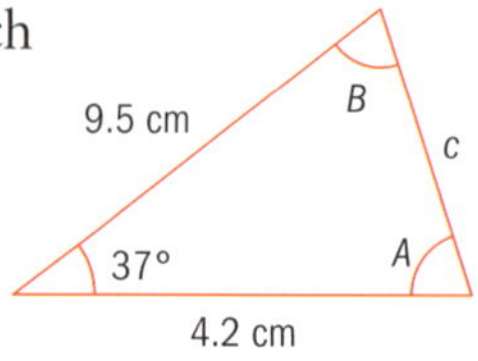

$c^2 = a^2 + b^2 - 2ab\cos C$

$c^2 = 9.5^2 + 4.2^2 - 2 \times 9.5 \times 4.2 \times \cos 37$

$c^2 = 44.1589...$

$c = 6.65$ cm (2 d.p.)

Skills check:

1. A right-angled triangle has side lengths 6.3 cm and 20.4 cm. The side of length 20.4 cm is the hypotenuse. Calculate the length of the missing side.

2. Calculate the length of BC.

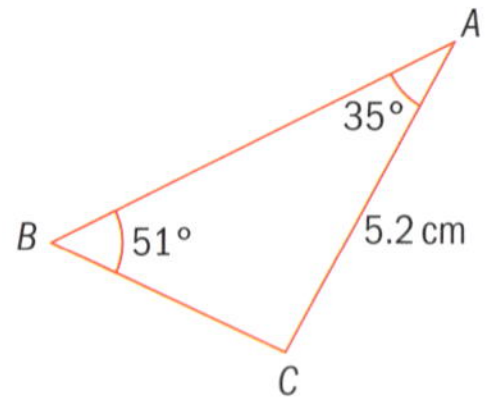

3. Find angle θ.

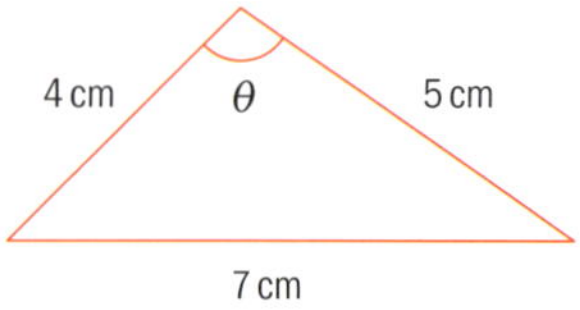

5.1 Three forces acting at a point

When three forces acting at a point are in equilibrium, then the forces can be represented by the sides of a triangle. Each force is proportional to the sine of the other angles and the resultant force is zero.

Recall the sine rule:
$$\frac{C}{\sin c} = \frac{B}{\sin b} = \frac{A}{\sin a}$$

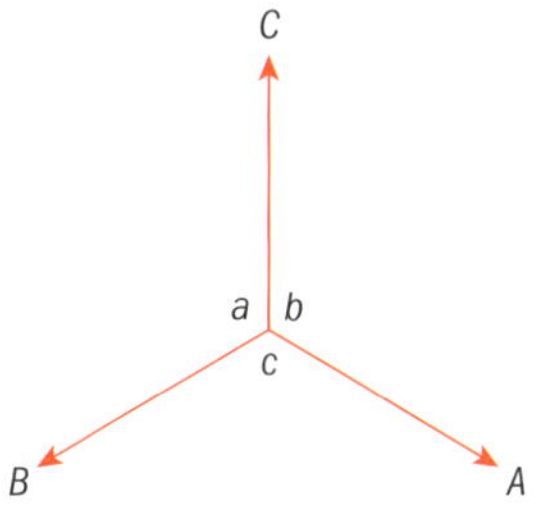

A body, and the forces acting upon it, is in equilibrium if the body is at rest or is moving with constant velocity.

Let us assume that the sum of the forces acting on a particle is zero. We can say that these forces are in equilibrium. Similarly, if the particle is at rest then we say that the forces acting are in equilibrium.

From the above diagram we can represent the forces on a triangle *XYZ*, as shown below.

Since the three forces are in equilibrium, their resultant is zero and hence the vector diagram, drawn nose-to-tail, is a closed figure – **the triangle of forces.**

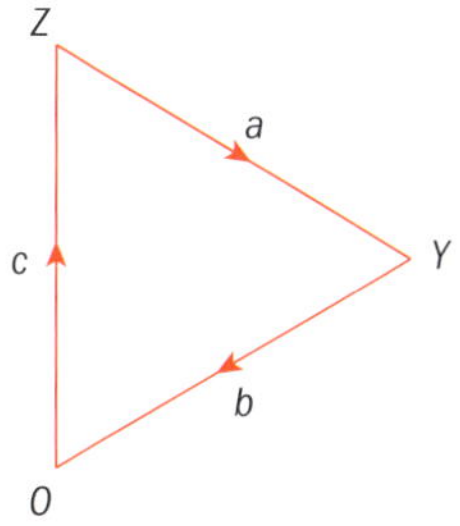

In the examples that follow, some terminology which we met in Chapter 4 is used in the context of mathematical modelling:

Example 1

A mass of 10 kg is suspended in equilibrium by two light inextensible strings *A* and *B*, which make angles of 15° and 35° respectively to the horizontal, as shown in the diagram. Calculate the tension in the strings.

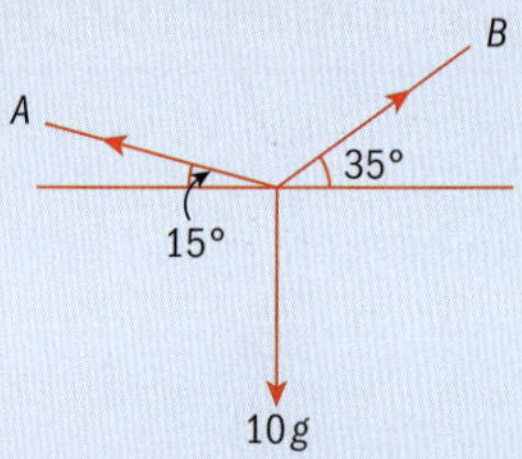

Take $g = 10\,\text{ms}^{-2}$.

The forces as shown in the diagram can be arranged to form the following triangle *XYZ*.

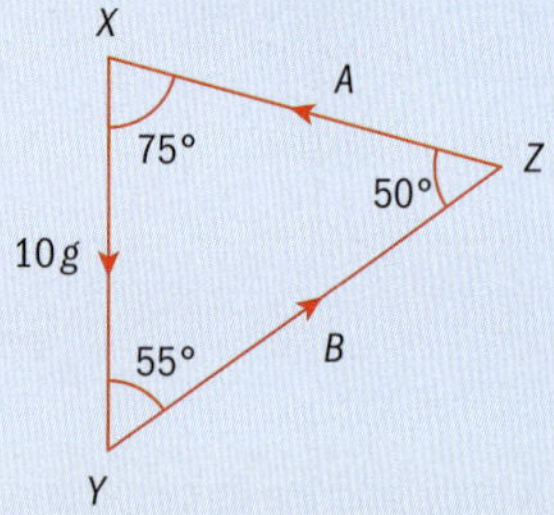

Three forces *A*, *B* and *C* acting at a point are in equilibrium, and can therefore be represented by the sides of a triangle *XYZ*.

$$\frac{A}{\sin 55°} = \frac{B}{\sin 75°} = \frac{10g}{\sin 50°}$$

$$A = \frac{\sin 125° \times 10g}{\sin 130°}$$

$A = 106.9\,\text{N}$

$$B = \frac{\sin 105° \times 10g}{\sin 130°}$$

$B = 126.1\,\text{N}$

An alternative approach would be to use the following:

$-A\cos 15 + B\cos 35 = 0 \quad (1)$

$A\sin 15 + B\sin 35 = 10g \quad (2)$

Solving (1)

$$A = \frac{B\cos 35}{\cos 15} \quad (3)$$

substitute (3) into (2)

$$\frac{B\cos 35}{\cos 15} \times \sin 15 + B\sin 35 = 10g$$

$$B = \frac{100}{\sin 35 + \tan 15 \times \cos 35}$$

$B = 126.1\,\text{N}$

Substitute B into (3)

$A = 106.9\,\text{N}$

Alternatively, once either B, or A has been found we can use the cosine rule to find the other side. For the following example, B (126.1N) has been found first of all.

$A^2 = 100^2 + 126.1^2 - 2 \times 100 \times 126.1 \times \cos 55$

$A = 106.9\,\text{N}$

Example 2

The system of forces shown is in equilibrium. Find X and θ.

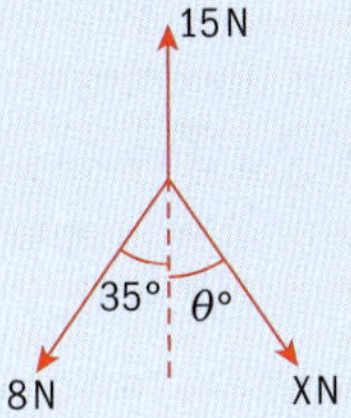

We can represent the system of forces using the triangle ABC:

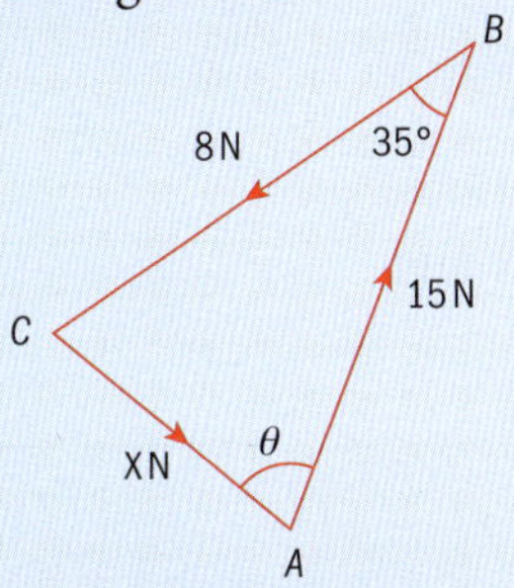

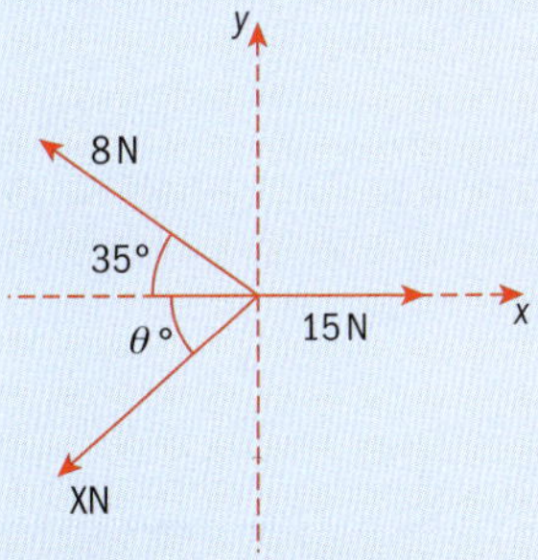

Resolving along the x-axis

$8\cos 35 + X\cos\theta = 15$ (1)

Resolving along the y-axis

$8\cos 35 + X\sin\theta$ (2)

Re-arranging (1)

$$X = \frac{15 - 8\cos 35}{\cos\theta} \quad (3)$$

Substitute (3) into (2)

$$8\sin 35 = \sin\theta\left(\frac{15 - 8\cos 35}{\cos\theta}\right)$$

$$\tan\theta = \frac{8\sin 35}{15 - 8\cos 35}$$

$\theta = 28.5°$

Substitute $\theta = 28.5°$ into (2)

$X = 9.61\text{ N}$

To find X, we use the cosine rule:

$X^2 = 15^2 + 8^2 - 2 \times 15 \times 8 \times \cos 35°$

$X^2 = 92.4035 \ldots$

$X = 9.61\text{ N}$

To find θ, we use Lami's theorem and the sine rule:

$$\frac{9.61}{\sin 35°} = \frac{8}{\sin\theta}$$

$$\sin\theta = \frac{8\sin 35°}{4.8}$$

$\theta = 28.5°$

We will now take the system of forces and plot on to an x and y axis, as shown in the diagram opposite.

An alternative approach is to use the fact that, for a system of forces in equilibrium (not necessarily three forces), the sum of the components in any direction is zero. It is necessary to choose two perpendicular directions and resolve forces in each of these directions. In the previous example we could choose to resolve in the direction of the 15 N force and perpendicular to this. This would give:

$$X\sin\theta = 8\sin 35 \text{ and } X\cos\theta + 8\cos 35 = 15$$

i.e. $X\cos\theta = 15 - 8\cos 35$

$$\tan\theta = \frac{(8\sin 35)}{(15 - 8\cos 35)}$$

giving $\theta = 28.5°$, $X = 9.61$, as before.

In the following example the same method is used, but this time the directions chosen when considering the particle on the plane are parallel to the plane and perpendicular to the plane.

Example 3

A light inextensible string passes over a smooth pulley fixed at the top of a smooth plane inclined at 30° to the horizontal. A particle of mass 2 kg is attached and hangs freely. A particle of mass m kg is attached to the other end of the string and rests in equilibrium on the surface of the plane. Calculate the normal reaction R between the mass m and the plane, the tension T in the string, and the value of m.

Take $g = 10\,\text{ms}^{-2}$

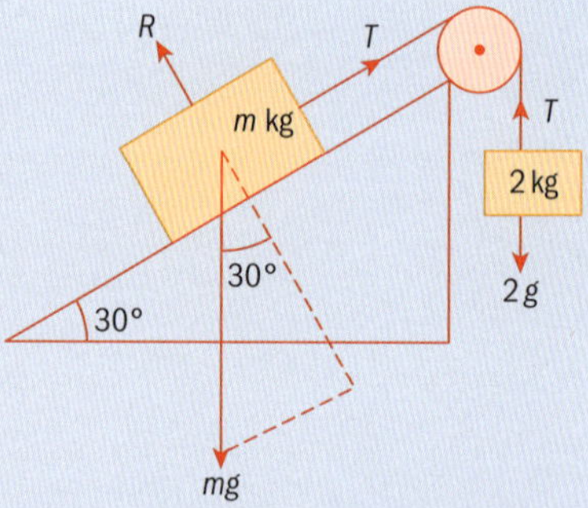

- **Resolving vertically:**

  For the 2 kg particle,

  $T = 19.6\,\text{N}$

- **Resolving parallel to the surface of the slope:**

  For particle of mass m kg,

  $T = 2g$

  $2g = mg\sin 30°$

  $m = 4$

- **Resolving perpendicular to the surface of the slope:**

  $R = mg\cos 30°$

  Substituting $m = 4$ into the equation above, we get:

  $R = 4g\cos 30°$

  $R = 4 \times 10 \times \cos 30°$

  $R = 34.6\,\text{N}$

Hint: $\sin 30° = 0.5$

The force R is called the **normal reaction** as it acts at right-angles to the plane.

Exercise 5.1

1. Each of the following systems of forces is in equilibrium. For each one, find the magnitude of the missing forces and the size of angle θ.

a)

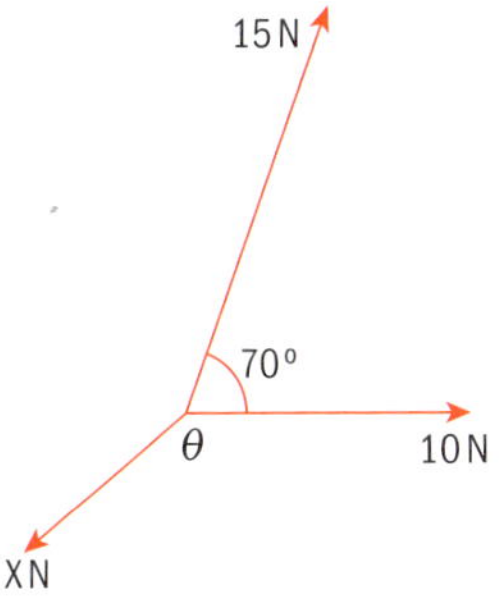

b)

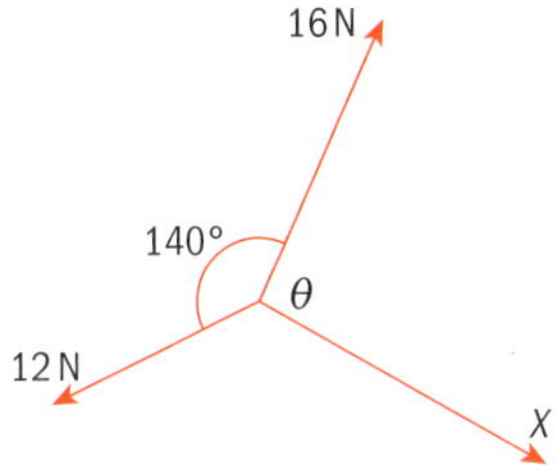

c)

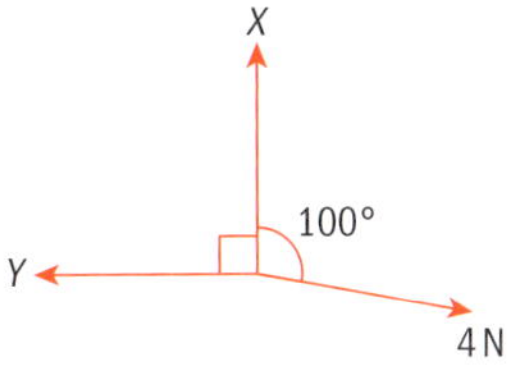

d)

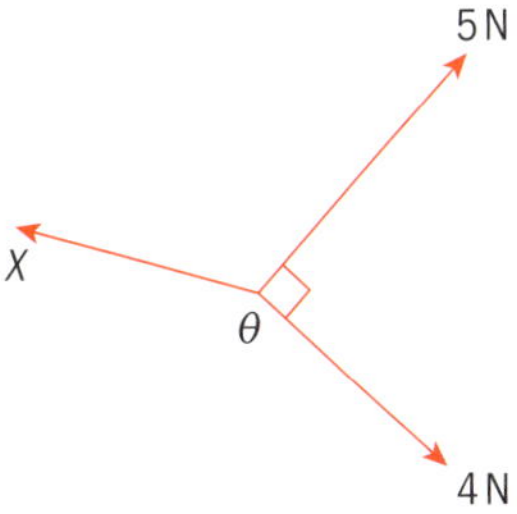

e)

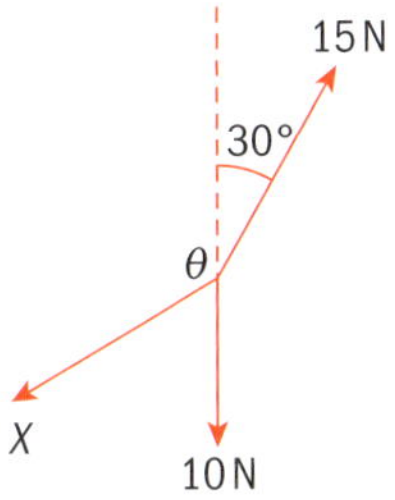

f)

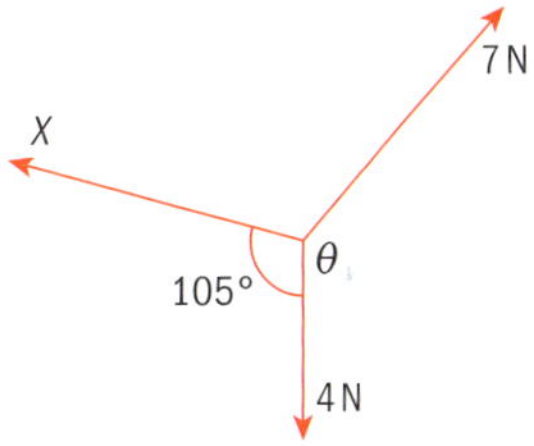

g)

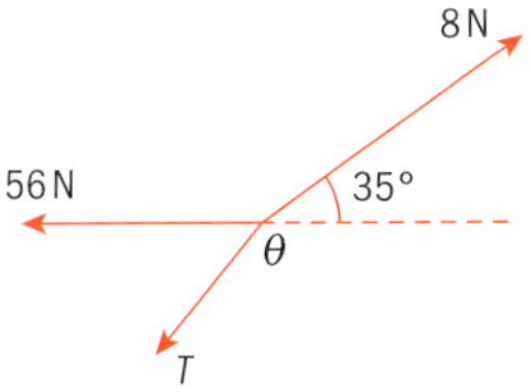

2. The forces acting at O cause the system to be in equilibrium. By resolving in two directions, find P and Q.

a)

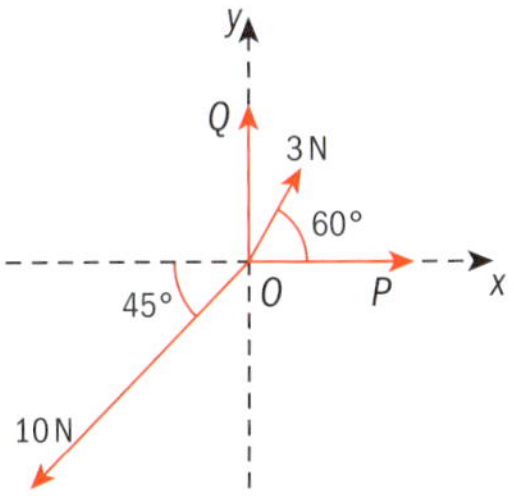

b)

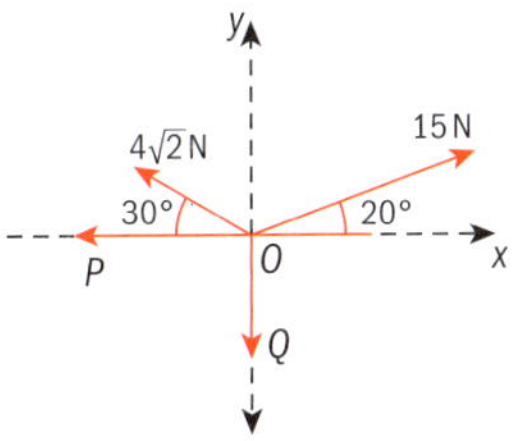

c)

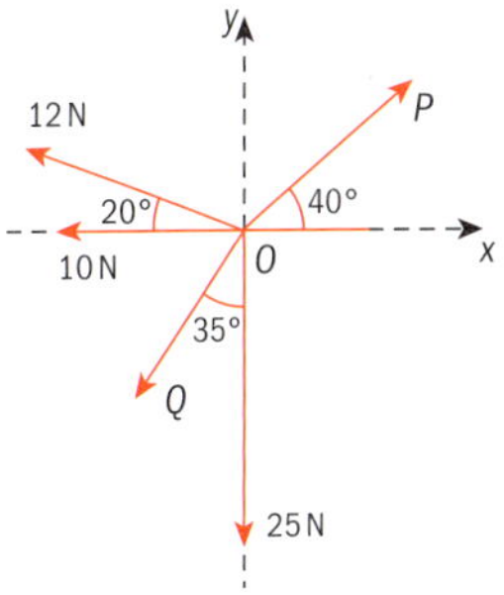

3. A bike of mass 10 kg is stationary on a smooth hill, which is inclined at 35° to the horizontal. When a force X is applied, the bike rests in equilibrium. The force is parallel to and acts up the plane. Calculate the force, X, and the normal reaction between the bike and the hill.

4. A crate of mass 12 kg is stationary on a smooth hill, which is inclined at 45° to the horizontal. Two forces are applied to the crate, which rests in equilibrium. One force is 50 N, which is parallel to and acts up the plane, and the other force X, is horizontal. Calculate the force X, and the normal reaction between the crate and the hill.

5. A light inextensible string passes over a smooth pulley fixed at the top of a smooth plane, inclined at 60° to the horizontal. A particle of mass 4 kg is attached and hangs freely. A mass m is attached to the other end of the string and rests in equilibrium on the surface of the plane.

   Calculate the normal reaction between the mass m and the plane, the tension in the string and the value of m.

6. A light inextensible string passes over a smooth pulley fixed at the top of a smooth plane, inclined at 60° to the horizontal. A force of 5 N at 30° to the surface of the slope is applied and is acting down the slope. A particle of mass 4 kg is attached and hangs freely. A mass m is attached to the other end of the string and rests in equilibrium on the surface of the plane.

   Calculate the normal reaction between the mass m and the plane, the tension in the string and the value of m.

Summary exercise 5

1. Given that the following set of forces are in equilibrium, find the missing values which are given as letters.

   a) $(3\mathbf{i} + 2\mathbf{j})$ N, $(5\mathbf{i} + 4\mathbf{j})$ N, $(a\mathbf{i} + b\mathbf{j})$ N

   b) $(7\mathbf{i} + \mathbf{j})$ N, $(-2\mathbf{i} - 6\mathbf{j})$ N, $(c\mathbf{i} + d\mathbf{j})$ N

   c) $(\mathbf{j})$ N, $(\mathbf{i})$ N, $(2\mathbf{i} + 3\mathbf{j})$ N, $(e\mathbf{i} + f\mathbf{j})$ N

   d) $\left(2\sqrt{3}\mathbf{i} - 2\mathbf{j}\right)$ N, $\left(-2\sqrt{3}\mathbf{i} - 5\mathbf{j}\right)$ N, $-\left(5\mathbf{i} - \sqrt{3}\mathbf{j}\right)$ N, $+(g\mathbf{i} - h\mathbf{j})$

2. Each of the following system of forces are in equilibrium. Find the magnitude of the missing force X.

   a)

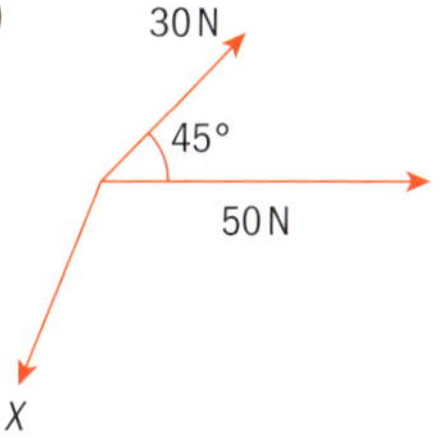

   b)

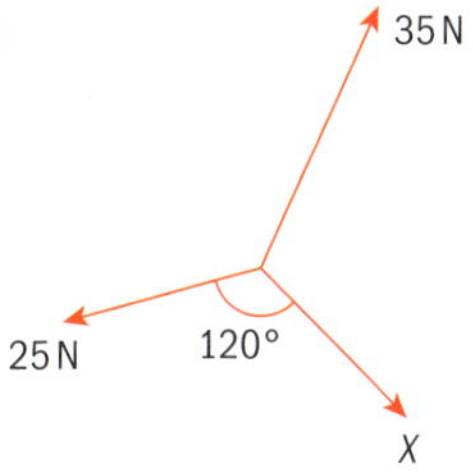

   c)

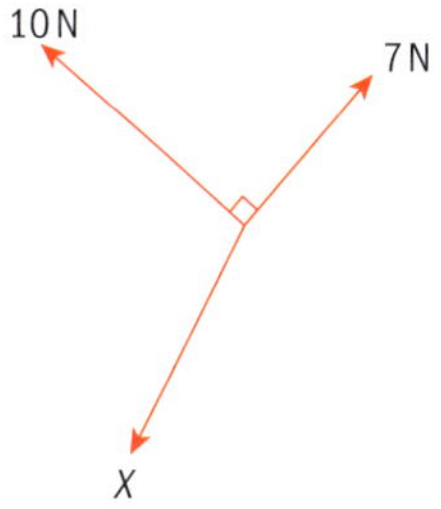

   d)

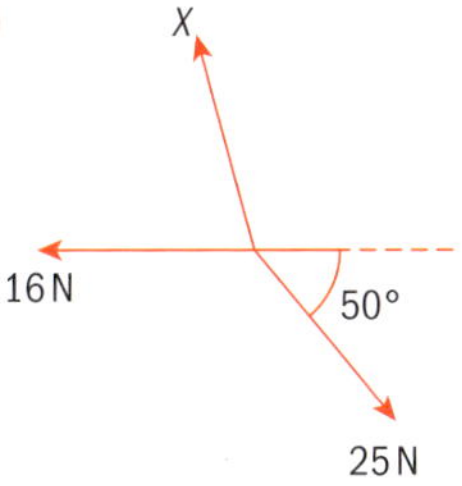

   e)

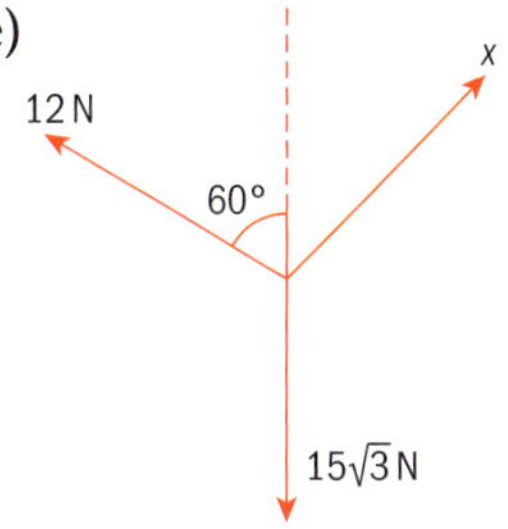

   f)

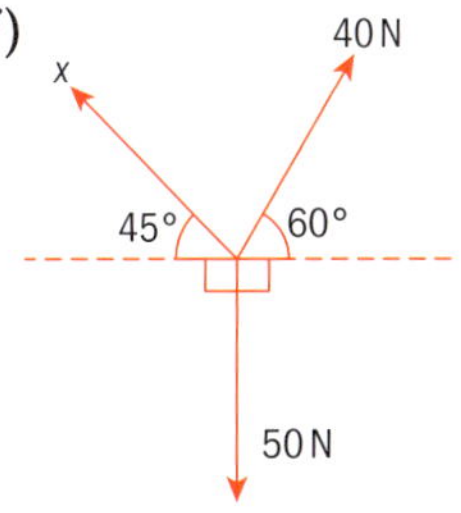

3. The forces acting at O causes the system to be in equilibrium. By resolving in two directions, find A and B.

   a)

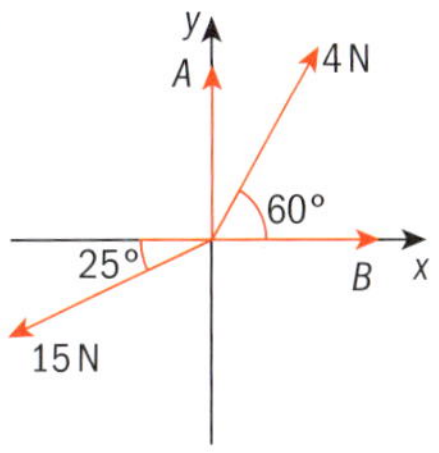

   b)

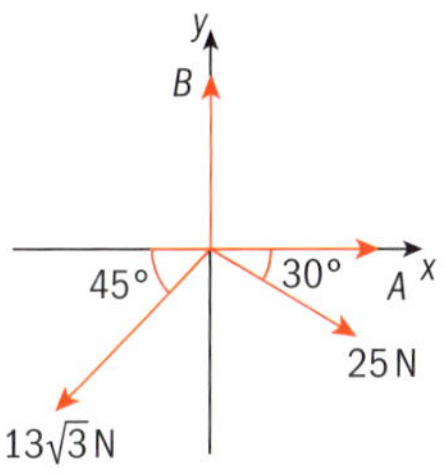

   c)

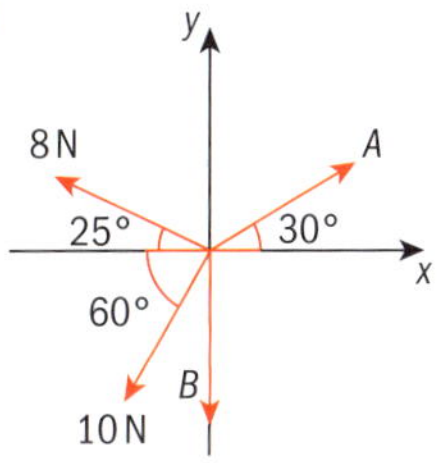

   d)

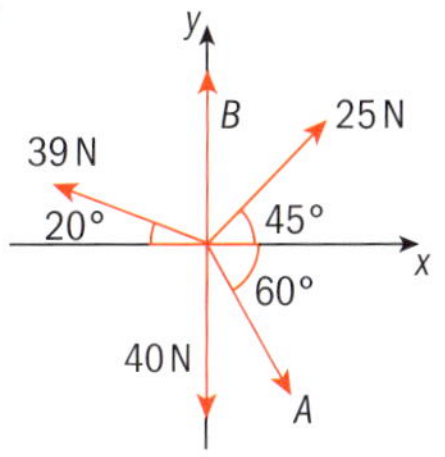

   e)

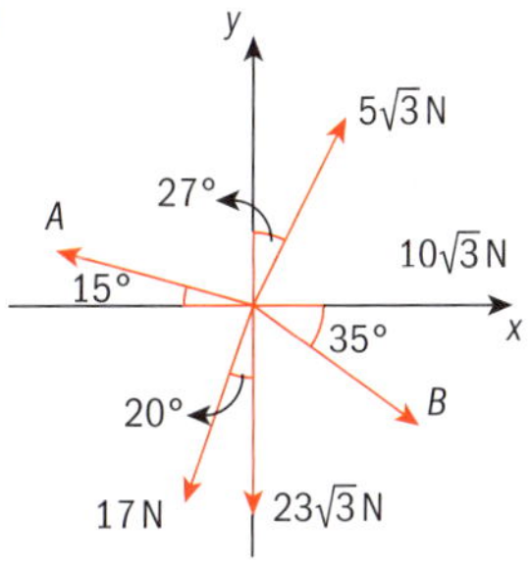

4. During a mountain stage of the Tour de France, a bike of mass 6.8 kg is stationary and on a hill, which is inclined at 6.2°. This type of climb is also known as a category four climb.
   When a force X is applied, the bike rests in equilibrium. The force is parallel and acts up the hill. Calculate the force X and the normal reaction between the bike and the hill.

5. During the Tour de France, a crate containing one of the team's equipment, weighing 300 kg, is stationary on a smooth hill which is inclined at 5.7° to the horizontal. Two forces are applied to the crate, which rests in equilibrium. One force is 650 N, which is parallel to and acts up the hill. The other force Y is horizontal. Calculate force Y and the normal reaction between the crate and the hill.

6. A light inextensible string passes over a smooth pulley fixed at the top of a smooth plane, inclined at 35° to the horizontal. A particle of mass 3.2 kg is attached and hangs freely. A mass m is attached to the other end of the string and rests in equilibrium on the surface of the plane. Calculate the normal reaction between the mass m and the plane, the tension in the string and the value of m.

7. A light inextensible string passes over a smooth pulley fixed at the top of a smooth plane, inclined at 45° to the horizontal. A force of 12 N at 25° to the surface of the slope is applied and is acting down the slope. A particle of mass $6\sqrt{3}$ kg is attached and hangs freely. A mass m is attached to the other end of the string and rests in equilibrium on the surface of the plane. Calculate the normal reaction between the mass m and the plane, the tension in the string and the value of m.

8. The following diagram shows a particle in equilibrium under the forces shown.

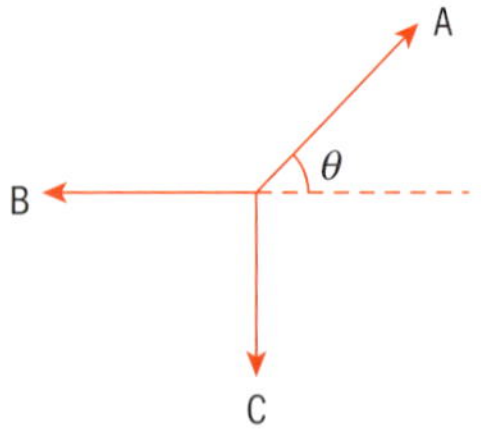

Show that C = B tan θ.

9. The following diagram shows a particle in equilibrium under the forces shown.

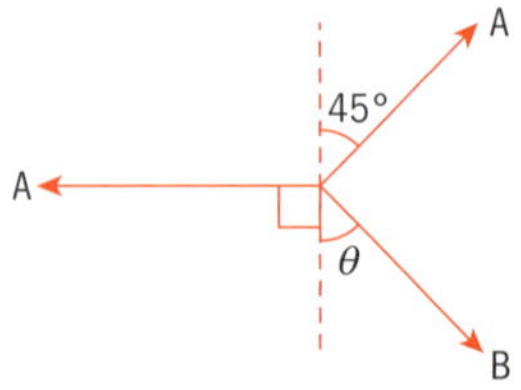

Show that $\tan\theta = \sqrt{2} - 1$.

EXAM-STYLE QUESTIONS

10.

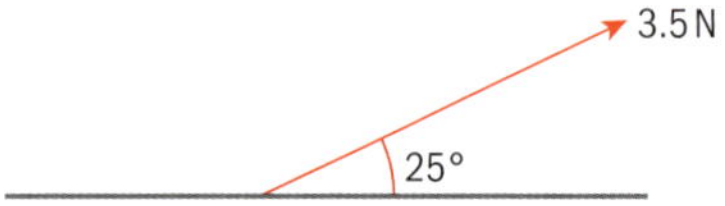

A shower curtain rail ring of mass 400 g is threaded on a rough curtain rail, which is fixed horizontally. Given that the ring is in equilibrium, acted on by a force of 3.5 N pulling upwards at 25° to the horizontal, find the normal component of the contact force acting on the rail ring.

11. A crate of mass 18.3 kg is at rest on a rough inclined plane at 8° to the horizontal. A force acts on the crate in a direction up the plane parallel to a line of greatest slope. When the magnitude of the force is 5A N the crate is on the point of sliding down the plane, and when the magnitude of the force is 12A N the crate is on the point of moving up the plane. Find the value of A.

12. 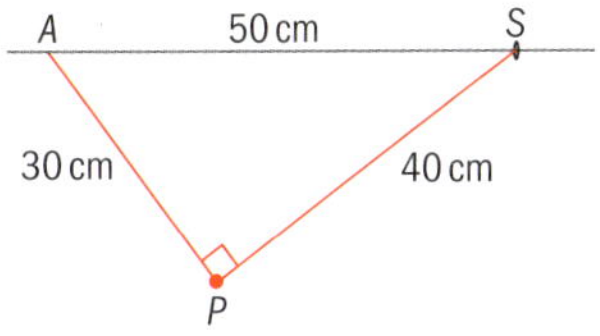

A particle P of weight 5 N is attached to one end of each of two light inextensible strings of lengths 30 cm and 40 cm. The other end of the shorter string is attached to a fixed point A of a rough rod which is fixed horizontally. A small ring S of weight W N is attached to the other end of the longer string and is threaded on to the rod. The system is in equilibrium with the strings taut and $AS = 50$ cm (see diagram).

i) By resolving the forces acting on P in the direction of PS, or otherwise, find the tension in the longer string. [3]

ii) Find the magnitude of the frictional force acting on S. [2]

iii) Given that the coefficient of friction between S and the rod is 0.75, and that S is in limiting equilibrium, find the value of W. [3]

Cambridge International AS and A Level Mathematics 9709, Paper 41 Q4 October/November 2009

Chapter summary

If three forces acting at a point are in equilibrium, then the forces can be represented by the sides of a triangle.

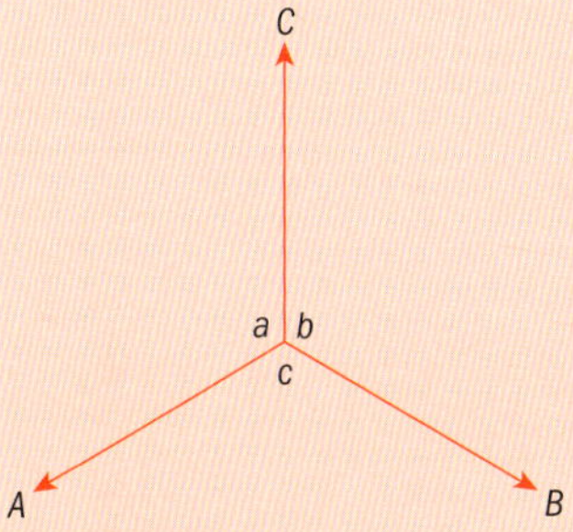

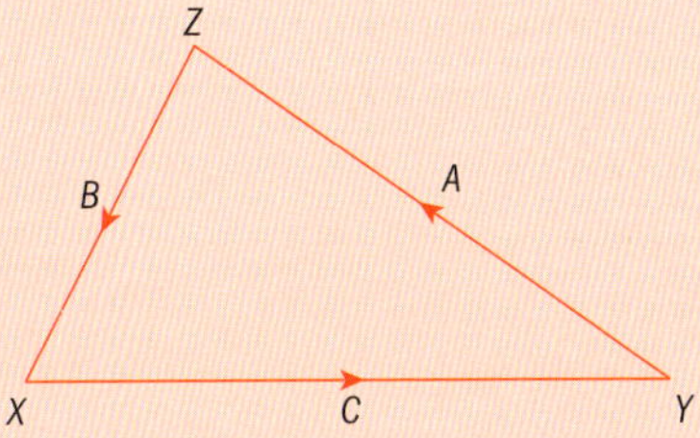

- Sine rule:

  $$\frac{a}{\sin A} = \frac{b}{\sin B} = \frac{c}{\sin C}$$

- Cosine rule:

  $a^2 = b^2 + c^2 - 2bc \cos A$

- Component of weight parallel to an inclined plane is $mg \sin \theta$.

  $F = mg \sin \theta$

- Component of weight perpendicular to an inclined plane is $mg \cos \theta$.

  $F = mg \cos \theta$

6 Friction

Friction is a force that opposes the motion between two surfaces in contact, and is encountered when an object slides on a surface. A **smooth** surface, for example a pane of glass, offers so little frictional resistance to the motion of the object sliding across it that the friction can be ignored.

A surface which is not smooth is said to be **rough**, such as the road shown on the left.

As most real objects experience friction it is an extremely important component in most applications of mechanics.

Objectives

- Understand that a contact force between two surfaces can be represented by two components, the normal component and the frictional component.
- Use the model of a 'smooth' contact, and understand the limitations of this model.
- Understand the concepts of limiting friction and limiting equilibrium; recall the definition of coefficient of friction, and use the relationship $F = \mu R$ or $F \le \mu R$, as appropriate.

Before you start

You should know how to:

1. Use trigonometric ratios to find missing angles.

$$\sin\theta = \frac{opposite}{hypotenuse};\ \cos\theta = \frac{adjacent}{hypotenuse};$$

$$\tan\theta = \frac{opposite}{adjacent}$$

e.g.

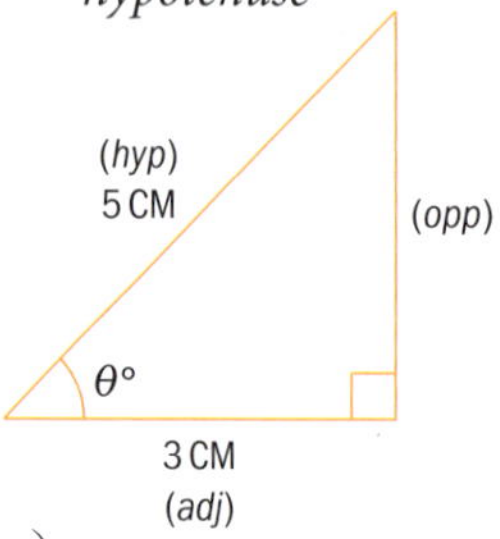

$$\cos\theta = \frac{adj}{hyp}$$

$$\cos\theta = \frac{3}{5} \Rightarrow \theta = \cos^{-1}\left(\frac{3}{5}\right) = 53.13° \text{ (2d.p.)}$$

Use trigonometric ratios to find missing sides.

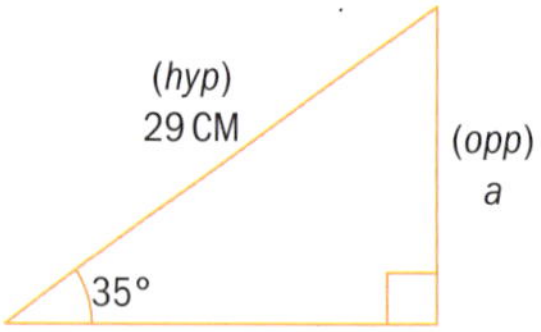

$$\frac{29}{\sin 90°} = \frac{a}{\sin 35°}$$

$$a = \frac{29 \sin 35°}{\sin 90°}$$

$$a = 16.63°$$

Skills check:

1. Find the missing angles in each of the following triangles.
   a) In triangle ABC, AC = 34 cm, AB = 30 cm and BC = 16 cm. Angle ABC is right-angles. Find cosA and angle CAB.
   b) In triangle PQR, PQ = 18 cm, PR = 24 cm and RQ = 30 cm. Angle RPQ is right-angled. Find cosQ and angle PQR.
   c) In the following right-angled triangle XYZ, XY = 15 cm and YZ = 20 cm. Find the size of the longest side. Find angle ZXY.

2. Re-arrange the following making a the subject.

$$v^2 = u^2 + 2as$$
$$a = \frac{v^2 - u^2}{2s}$$

3. Substituting into formulae.

e.g. Substitute the following values into the equation to find v.

$v = u + at$, when $a = 10$, $t = 3$ and $u = 5$.

$v = 5 + 10(3)$

$v = 35$

2. With the following equations, make the letter in brackets the subject.

a) $v = u + at$ (a)

b) $s = ut + \frac{1}{2}at^2$ (a)

c) $p^2 = \sqrt{\frac{g^3}{t}}$ (t)

3. For the following, substitute values into the formulae and solve.

a) $v^2 = u^2 + 2as$; $u = 3$, $a = 5$, $s = 3$

b) $s = ut + \frac{1}{2}at^2$; $u = 0.5$, $t = 3$, $a = 10$

c) $p^2 = \sqrt{\frac{g^3}{t}}$; $g = \frac{2}{5}$, $t = \frac{3}{7}$

Frictional forces

Imagine a horizontal force X being applied to a heavy object standing on a rough horizontal floor. As X increases, the frictional force also increases, so that they have the same magnitude but in opposite directions, until the frictional force reaches its maximum value, which is denoted by F_{max}.
An object will begin to move when the force being applied is greater than the frictional force present. On the point of moving, the object is said to be in the state of **limiting equilibrium** and

$$F = \mu R$$

where R is the **normal reaction** and μ is the **coefficient of friction** (which usually lies between 0 and 1). For perfectly smooth surfaces the coefficient of friction would be 0.

The Greek letter μ is pronounced “mu”.

Friction can be regarded as being the ratio of the weight of an object being moved along a surface and the force that maintains contact between the object and the surface. This is given as

$$F_{max} = \mu R$$

6.1 Rough horizontal surfaces

Example 1

Using the diagram, calculate the maximum frictional force which can act when a child's sledge rests on a rough horizontal surface when the coefficient between the surfaces is

a) 0.1 **b)** 0.5 **c)** 0.8

Take g as $10\,ms^{-2}$

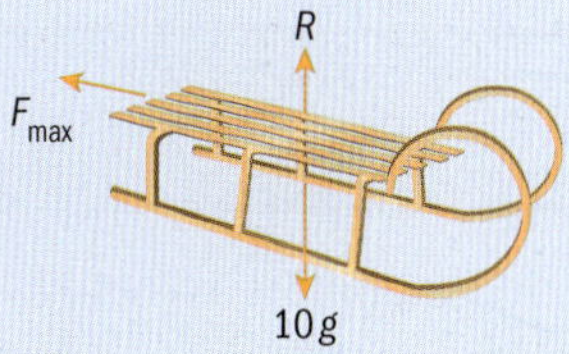

▶ Continued on the next page

As the sledge is at rest, there is no motion perpendicular to the ground.

Resolving vertically:

$R = 10g$
$R = 10 \times 10 = 100\,\text{N}$

a) $F_{max} = \mu R$ (maximum frictional force)
$F_{max} = \mu R$
$= 0.1 \times 100$
$= 10\,\text{N}$

b) $F_{max} = \mu R$
$= 0.5 \times 100$
$= 50\,\text{N}$

c) $F_{max} = \mu R$
$= 0.8 \times 100$
$= 80\,\text{N}$

In Example 1, we looked at an object at rest. In Example 2, we will explore what happens when parallel forces are applied to an object on a rough horizontal surface.

Example 2

A brick of mass 2 kg is at rest on a rough horizontal surface. The coefficient of friction between the brick and the surface is 0.9.

Calculate the frictional force acting on the brick when a horizontal force X is applied to the brick, and its magnitude is

a) 10 N **b)** 15 N **c)** 35 N

Calculate the magnitude of any acceleration that takes place due to motion.

Take g as $10\,\text{ms}^{-2}$.

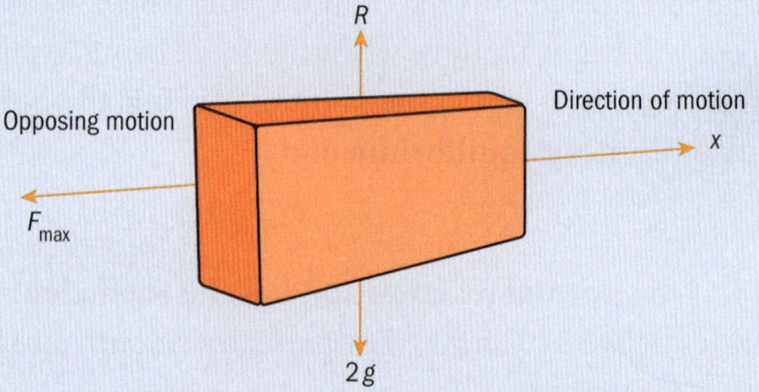

$F_{max} = \mu R$
$= 0.9 \times 2g$
$= 18\,\text{N}$

a) $F_{max} > 10\,\text{N}$, which means that no motion can take place.

b) $F_{max} > 15\,\text{N}$, which means that no motion can take place.

c) $F_{max} < 35\,\text{N}$, which means that motion will take place along the rough surface.

To find its acceleration we will use $F = ma$

$(X - F_{max}) = ma$
$(35 - 18) = 2 \times a$
$a = 8.5\,\text{ms}^{-2}$

Look back to Chapter 4 where Newton's second law, $F = ma$, was introduced.

Next, in Example 3, we will look at what happens when parallel forces are applied to an object on a rough horizontal surface, and the forces are not parallel to the motion of the object.

Example 3

A 12 kg basket containing fruit and vegetables from the local market is left on a rough horizontal floor. The coefficient of friction between the basket and the rough floor is 0.65. The basket is about to move when a force *XS* is applied at an angle θ to the horizontal. Calculate the magnitude of *X* when

a) $\theta = 0$

b) $\theta = 15$

c) $\theta = 45$

d) $\theta = 70$.

Take g as $10\,\text{ms}^{-2}$.

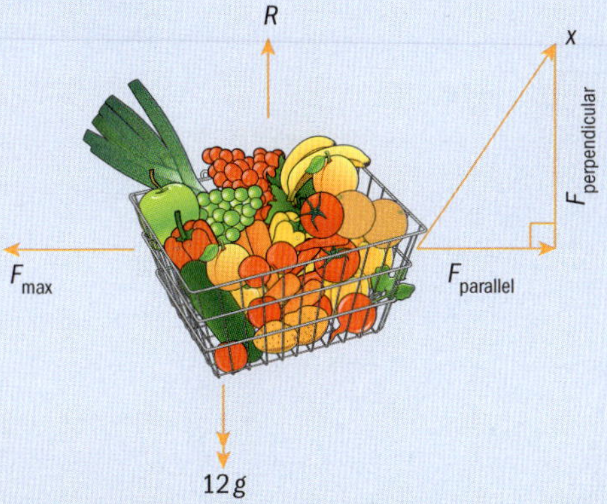

a) Resolving vertically:

$R = mg$

$\quad = 12g = 12 \times 10$

$\quad = 120\,\text{N}$

In the state of limiting equilibrium, $F_{max} = X = \mu R$

$X = 0.65 \times 120$

$X = 78\,\text{N}$

This means that for motion to occur, the force being applied must exceed 78 N.

b) From the above diagram, we can work out the forces perpendicular and parallel to the floor.

$\sin 15° = \frac{F_{perpendicular}}{X}$, so $F_{perpendicular} = X \sin 15°$

$F_{parallel} = X \cos 15°$

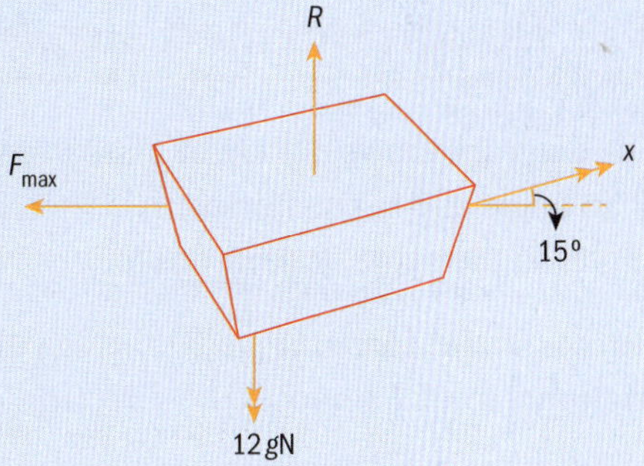

Resolving vertically:

$R + X \sin 15° = 12g$

$R = 12g - X \sin 15°$

At the time when limiting equilibrium takes place:

$F_{parallel} = X \cos 15° = F_{max} = \mu R$

$X \cos 15° = 0.65\,(12g - X \sin 15°)$

$X(\cos 15° + 0.65 \sin 15°) = 0.65 \times 12g$

$X = 68.77\,\text{N}$

For motion to take place, the force must exceed 68.77 N

Exercise 6.1

The following diagrams show a solid block of granite of mass 700 kg at rest. Given that the coefficient of friction between the block and the surface is 0.4, state whether the forces applied are enough to cause it to move.

1.

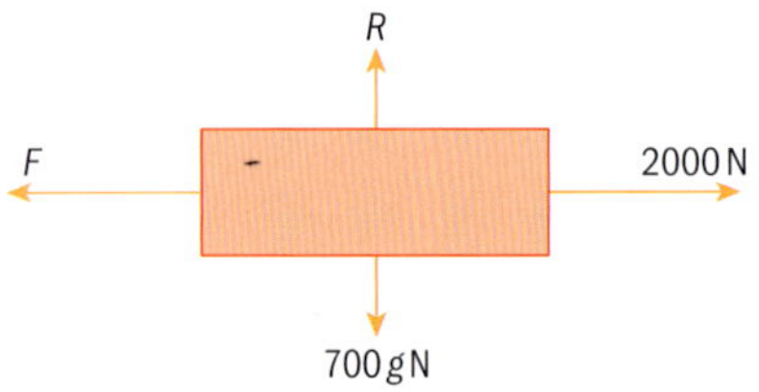

2.

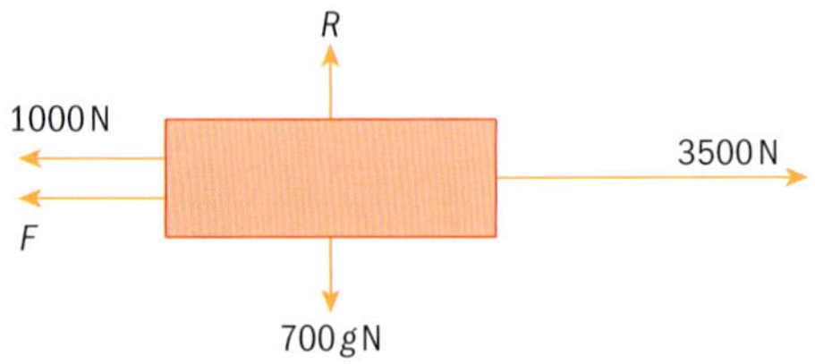

3.

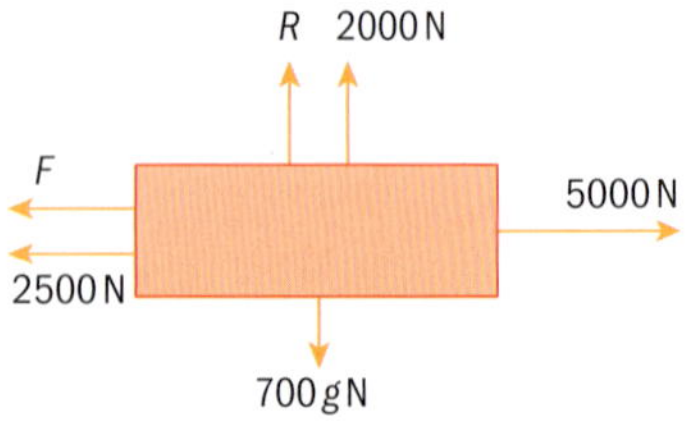

4.

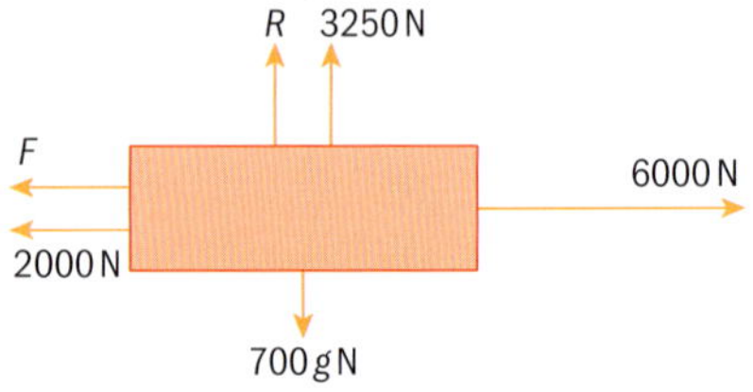

5.

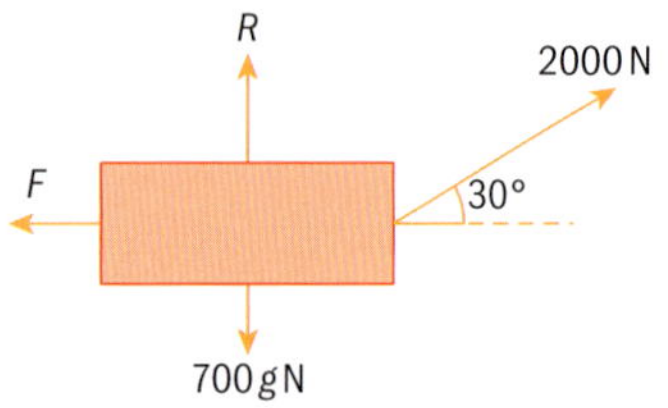

6.

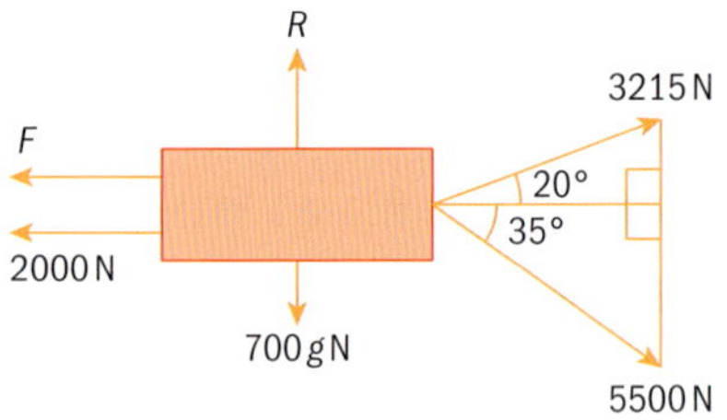

6.2 Rough inclined slope

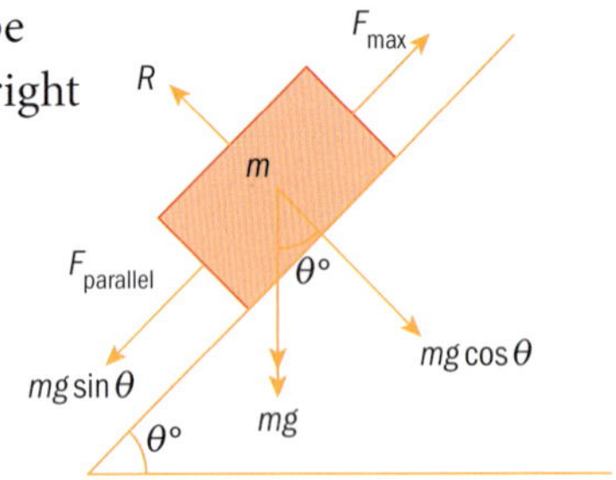

The diagram shows a body of mass m resting on a rough inclined slope at an angle of θ to the horizontal. There is no motion, so resolving at right angles to the plane, $R = mg\cos\theta$.

As the parallel force $mg\sin\theta$ is acting down the slope, the frictional force F opposes the direction of this force.

In the system's equilibrium state:

$$F_{max} = \mu R$$
$$= \mu mg\cos\theta$$

Example 4

A sledge of mass 10 kg rests in limiting equilibrium on a rough inclined slope, 30° to the horizontal. Find the coefficient of friction between the sledge and the slope.

Take g as $10\,\text{ms}^{-2}$.

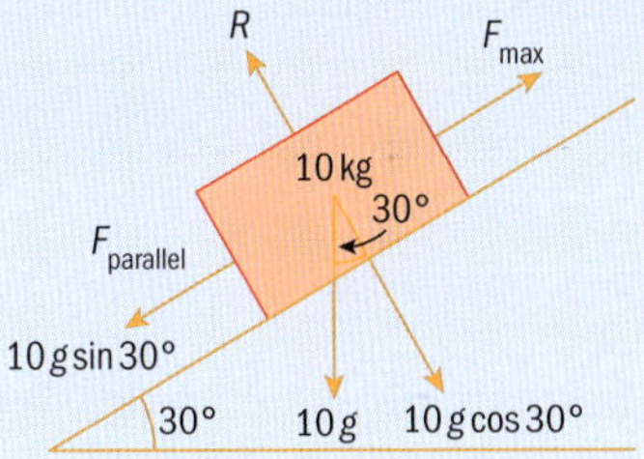

Resolving parallel to the slope:

$10g\sin 30° = \mu R$

Resolving perpendicular to the slope:

$R = 10g\cos 30°$

Substituting this value of R into the equation above, we have:

$10g\sin 30° = \mu R = \mu \times 10g\cos 30°$

$\mu = \dfrac{10g\sin 30°}{10g\cos 30°} = \tan 30°$

$\mu = \dfrac{\sqrt{3}}{3}$

Remember that $\sin 30° = \dfrac{1}{2}$ and $\cos 30° = \dfrac{\sqrt{3}}{2}$, and that $\dfrac{1}{\sqrt{3}}$ can also be written as $\dfrac{\sqrt{3}}{3}$.

Example 5

A van is carrying fruit from the local market. It breaks down on a steep hill inclined at 30° to the horizontal. Given that the van and fruit have a mass of 1500 kg and the coefficient of friction between the van's wheels and the slope is 0.56, find the force X parallel to the slope, which must be applied to the van to prevent motion down the slope.

Take g as $10\,\text{ms}^{-2}$.

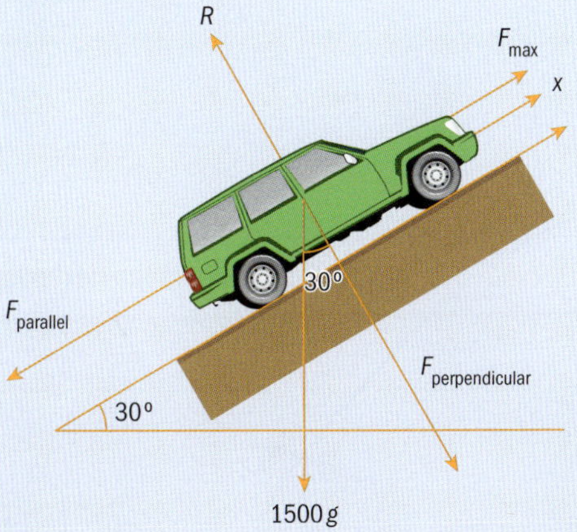

▶ Continued on the next page

As the van is prevented from moving down the slope, frictional force is in the opposite direction, opposing motion.

Resolving perpendicular to the slope:

$R = 1500g \times \cos 30°$
$R = 1500 \times 10 \times \cos 30° = 12\,990.38\text{ N}$

Resolving parallel to the slope:

$F_{max} + X = 1500g \times \sin 30°$

$X = 1500g \times \sin 30° - \mu R$

$X = 1500g \times \sin 30° - 0.56 \times 12\,990.38$

$X = 225.39\text{ N}$

$F_{max} = \mu R$

To prevent the van slipping down the slope, a force of 225.39 N must be applied.

Example 6

A trailer full of potatoes is at rest on a hill inclined at 55° to the horizontal.

The coefficient of friction between the trailer and the slope is $\frac{\sqrt{3}}{2}$.

Given that the trailer with potatoes has mass 3500 kg, what is the trailer's acceleration if a force of 47 000 N is applied up the hill along a line of greatest slope?

Take g as 10 ms^{-2}.

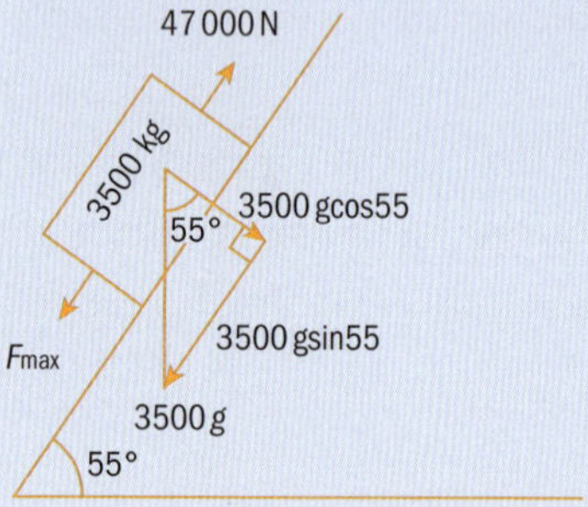

▶ Continued on the next page

Resolving perpendicular to the slope:

$R = 3500g \times \cos 55°$
$R = 3500 \times 10 \times \cos 55° = 20\,075.18$ N

If the trailer is moving up the hill, the frictional force, F_{max}, must be acting down the hill.

$F_{max} = \mu R = \frac{\sqrt{3}}{2} \times 20\,075.18$
$= 17\,385.62$ N

Also acting down the hill is the parallel force of $3500g \sin 55° = 28\,670.32$ N

The total force down the slope is $17\,385.62 + 28\,670.32 = 46\,055.94$ N

Since the force exerted on the trailer going up the hill is greater than the force opposing the trailer, then motion will occur.

Using $F = ma$,

$47000 - F_{max} - 3500g \sin 55° = 3500a$
$47000 - 17385.62 - 28670.32 = 3500a$

So the acceleration, $a = 0.27$ ms^{-2}

Exercise 6.2

1. A sledge of mass 15 kg is released from rest on a rough inclined plane. R is the normal reaction and F is the frictional force exerted on the body by the plane. For each of the following, calculate the magnitude of F and state whether the sledge will remain at rest or will move up or down the slope.

a)

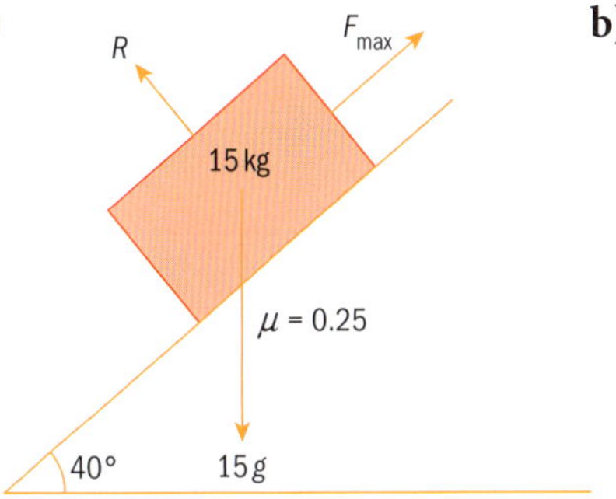

b)

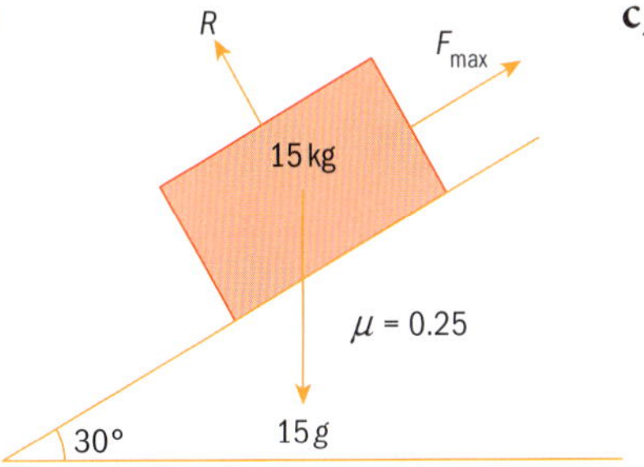

c)

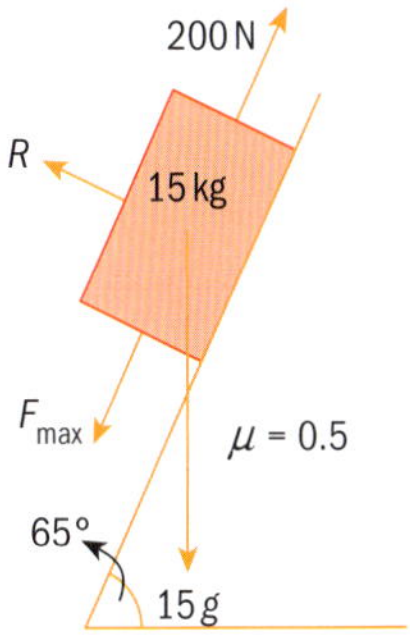

2. A skier of mass 92 kg is just on the point of being pulled up the ski slope by a ski lift. However, the skier is on part of the ground that has just thawed, making it difficult to move. Given that the angle of the slope is 55° to the horizontal, and the coefficient of friction is 0.86, find the magnitude of the force needed to keep the skier on the point of moving up the slope.

3. A crate containing large plastic water bottles is being delivered to houses in and around a city. The first delivery is on a hill of 17° inclined to the horizontal. Given that the van is pointing down the hill, the mass of the crate and bottles are 230 kg and the coefficient of friction between the crate and van is 0.5, what force must be applied in order for the crate to stop sliding further into the van?

4. A large block is sliding down a hill inclined at 45° to the horizontal with an acceleration of 0.5 ms^{-2}. Given that the block weighs 1500 kg, what is the coefficient of friction between the block and the hill.

5. A wooden block of mass 250 kg is being pulled up a hill by a car accelerating at 0.25 ms^{-2}. The hill is inclined at 30° to the horizontal. The force being applied to the wooden block is 2000 N at 35° to the slope as shown in the diagram. Find the value of μ.

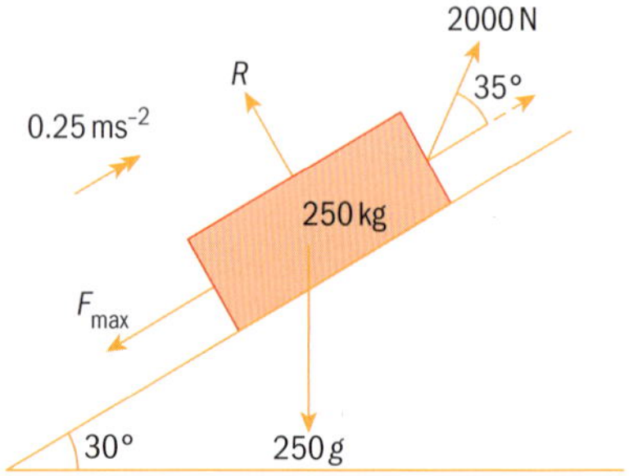

6. The diagram on the right shows a crate of mass 100 kg being pulled up an inclined hill at 30° to the horizontal. One man uses a force of 650 N to pull the crate in a motion parallel to the slope. Another man pulls with 175 N inclined at 20° parallel to the slope. If a force of 100 N is opposing this motion at 45° parallel to the slope, what is the coefficient of friction?

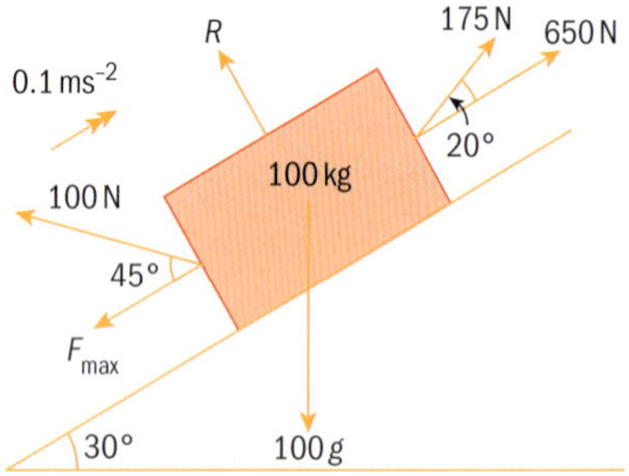

Summary exercise 6

1. A suitcase of mass 5 kg is released from rest at the top of a slippery slope which is fixed at 35° to the horizontal. The coefficient of friction between the suitcase and the slope is 0.15. Ignoring any other resistances, calculate the acceleration of the suitcase as it makes its way down the slope.

2. A block of mass 2 kg is placed on a rough surface, which is inclined at 40° to the horizontal. The block moves down the incline at 0.5 ms^{-2}. Find the coefficient of friction between the block and the surface.

3. A man is delivering a heavy box of 20 kg to a house on top of a hill, when all of a sudden he lets it slip and it slides down the hill. The hill is inclined at 25° to the horizontal and the coefficient of friction between the box and the hill is 0.65. Assuming constant acceleration and given that the initial speed of the box is 4 ms^{-1}, what distance does the box cover before it stops?

4. Two men are delivering a crate of mass 100 kg to a house on a hill which is inclined at 35° when it is dropped and slides down the hill. The coefficient of friction between the crate and surface is 0.8. A gust of wind suddenly appears and applies a force of 100 N to the crate parallel to, and facing up, the surface of the hill for a total of 5 seconds. During this time the crate covers a total distance of 32.5 m. Given that the acceleration is constant, what was the crate's initial speed?

5. A briefcase of mass 5 kg is at rest on a rough horizontal surface. The coefficient between the briefcase and the surface is 0.68. State whether the following forces being applied are enough to cause it to move.

   a) 20 N **b)** $11\sqrt{3}$ N **c)** $36\sqrt{6}$ N

   Calculate the magnitude of any acceleration that takes place due to its motion.

6. A basket containing bicycle components, from a local bike store is resting on the rough horizontal floor of the shop. The coefficient of friction between the floor and the basket is 0.27. The basket is just about to move when a force of X is applied at an angle of 0° to the horizontal. Calculate the magnitude of X when

   a) $\theta = 0°$ **b)** $\theta = 20\sqrt{3}°$

   c) $\theta = 35°$ **d)** 85°

7. The following diagrams show a block of wood of mass 65 kg at rest. Given that the coefficient of friction between the block and the surface is 0.47, state whether the forces applied are enough to cause it to move.

   a)

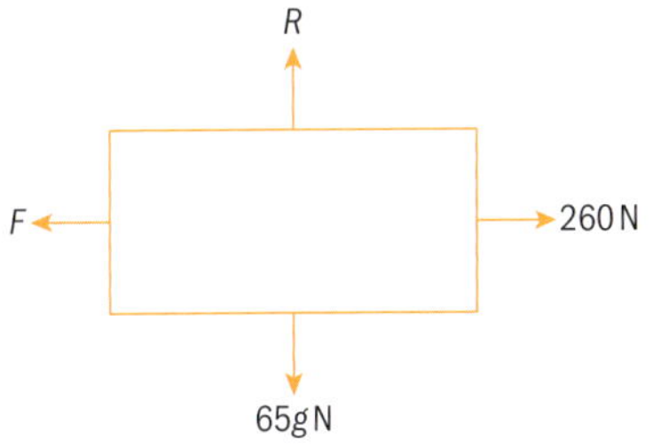

   b)

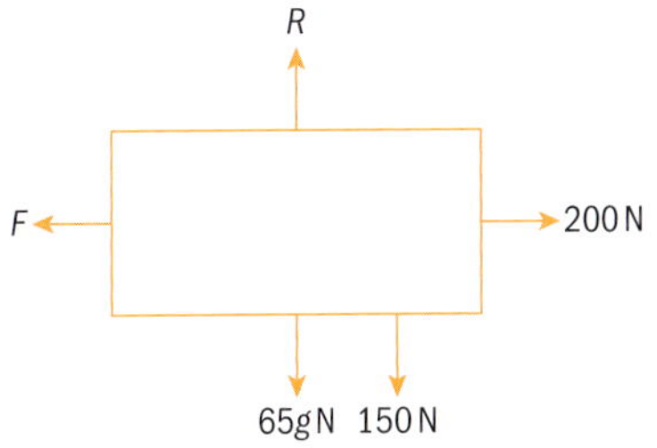

c) 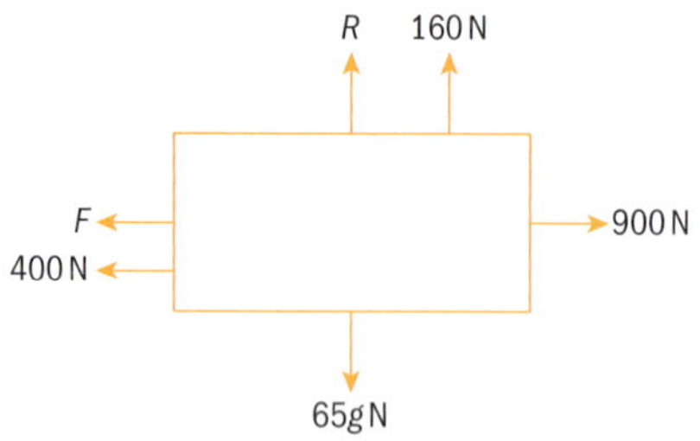

d) 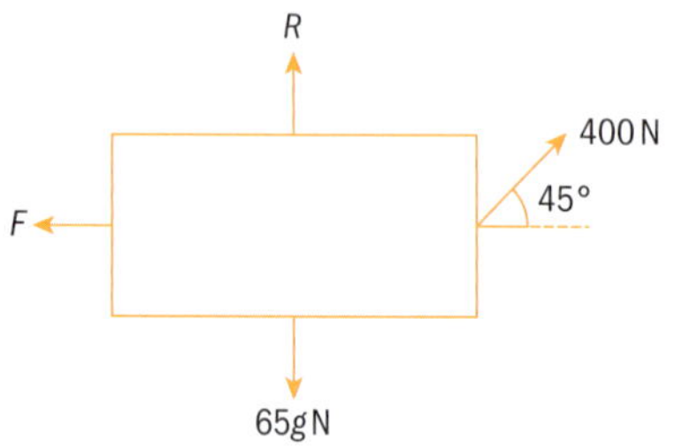

e) 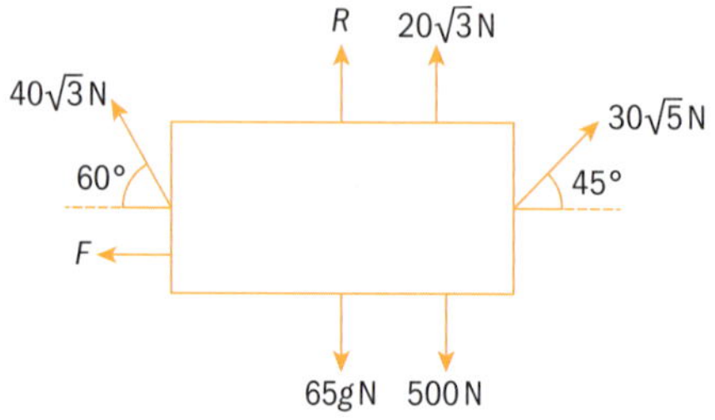

8. A fish tank of mass 60 kg rests in limiting equilibrium on a rough inclined slope at 35° to the horizontal. Find the coefficient of friction between the tank and the slope.

9. At the Tour de France, a trailer full of team bikes rests on a hill inclined at 6.1° to the horizontal. The coefficient of friction between the trailer and the slope is $\frac{\sqrt{5}}{4}$. Given that the trailer with bikes has a mass of 1068 kg, what is the trailer's acceleration if a force of 8000 N is applied uphill along a line of greatest slope?

10. A container full of boxes of cereal is sliding down a hill inclined at 25° to the horizontal with acceleration of 25 ms^{-2}. What is the coefficient of friction between the container and the hill?

11. The diagram shows a container load of bicycle parts of mass 80 kg being pulled up an inclined hill at 12.6° to the horizontal and the container is accelerating at $\frac{2}{\sqrt{3}}$ms^{-2}.

A small horse uses a force of 950 N to pull the crate in a motion parallel to the slope. To stop the container moving across the slope, another man pulls with a force of 210 N inclined at 30° parallel to the slope. What is the coefficient of friction between the container and the slope?

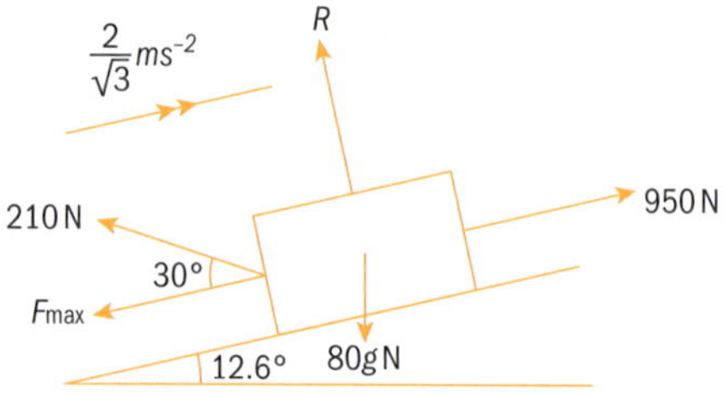

EXAM-STYLE QUESTIONS

12. A particle of mass 35 kg is in equilibrium on a plane that is inclined at an angle of θ° to the horizontal. The normal reaction force acting on the particle has magnitude 342 N. Find the

a) value of θ

b) the least possible value of the coefficient of friction.

13. A suitcase of mass 30 kg rests in limiting equilibrium on a horizontal. A force of magnitude 150 N acts on the suitcase at an angle of 25° to the upwards vertical. Find the coefficient of friction between the suitcase and the ground.

Chapter summary

- When frictional force reaches its maximum, $F_{max} = \mu R$
- On an inclined plane and perpendicular to the incline, Component of weight = $mg \cos \theta$
- On an inclined plane and parallel to the incline, Component of weight = $mg \sin \theta$
- Frictional force:

  F_{max} > resultant force in direction of motion implies that no motion will take place

  F_{max} < resultant force in direction of motion implies that motion will take place

Review exercise B

In these questions take the value of g as $10\,ms^{-2}$.

1. A box slides across a rough horizontal floor with a velocity of $2.4\,ms^{-1}$. The mass of the box is 1.2 kg and it comes to rest after sliding a distance of 7.5 m. Calculate the resistance force that is slowing the box down.
2. A rocket-propelled sled of mass 750 kg is moving across the desert floor powered by the rocket, which exerts a force of 4000 N. If the acceleration of the sled is $5\,ms^{-2}$, find the resistance acting on the sled. If the sled accelerates from rest for 10 seconds and then the power is shut off, calculate the distance that it will take the sled to come to rest.
3. A bucket weighing 3 kg is lowered down a well on a rope. If it is lowered with an acceleration of $0.5\,ms^{-2}$, find the tension in the rope. The bucket is filled with water weighing 4 kg. The bucket is pulled back up with an acceleration of $0.2\,ms^{-2}$. Find the tension in the rope on the way up.
4. A small block of mass 2 kg is being pulled up a rough plane by a string. The surface of the plane exerts a frictional force of 2.8 N on the block. The plane is inclined at an angle of 25° to the horizontal. If the tension in the string is 12.5 N, calculate the acceleration of the block.
5. A child's toy, weighing 0.8 kg is being pulled along the ground by a rope. The rope is being held above the ground so that it is at an angle of 15° to the horizontal. The tension in the rope is 1.5 N and the resistance exerted by the ground on the toy is 0.5 N. Calculate the acceleration of the toy.
6. A suitcase of mass 19.5 kg is sliding down a slope which is at an angle of 30° to the horizontal in the luggage handling area in an airport. If the suitcase is accelerating at $2\,ms^{-2}$, calculate the force of resistance that the surface of the slope exerts on the suitcase.
7. Two particles of mass 1.3 kg and 2.7 kg are connected by a light inextensible string, which hangs over a smooth pulley. Calculate the acceleration of the system when it moves under gravity.
8. 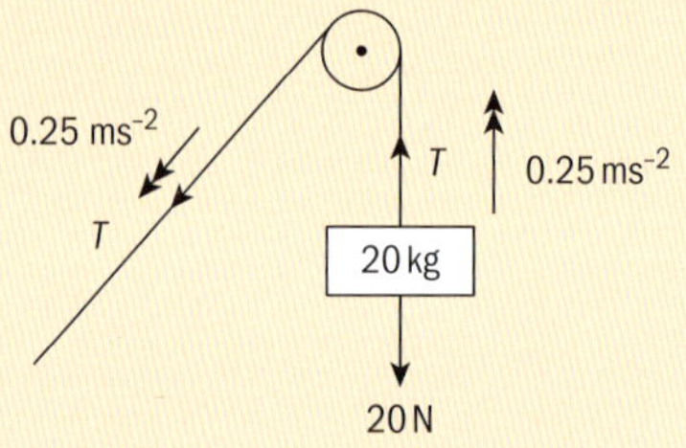

A heavy load is being hauled up on a building site by a rope attached to a pulley. The rope is being pulled at an angle of 40° to the vertical. If the acceleration of the load is $0.25\,ms^{-2}$, calculate the force with which the rope is being pulled, and the magnitude and direction of the reaction at the point where the pulley is fixed.

9. A small block of mass 2 kg rests on a horizontal table. It is connected by a light inextensible string that passes over a smooth pulley to another block of mass 3 kg which is hanging freely. If the surface of the table exerts a frictional force of 15 N, calculate the acceleration of the system and the tension in the string.
10. 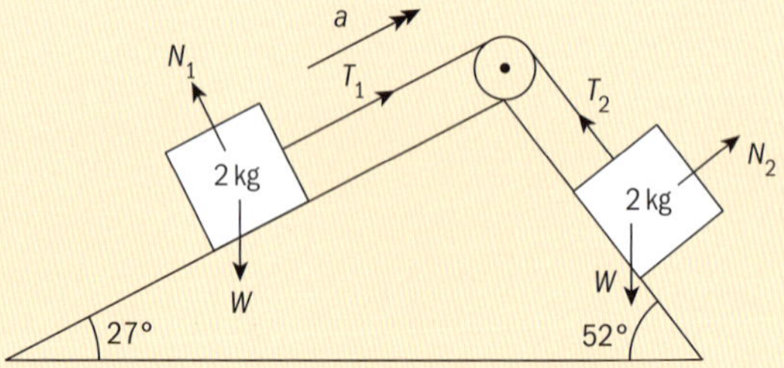

Two blocks of mass 2 kg are attached by a light inextensible string that passes over a smooth pulley. They hang either side of the pulley on smooth inclined planes that are at 27° and 52° to the horizontal respectively.

The system is released from rest. Calculate its resulting acceleration.

11. 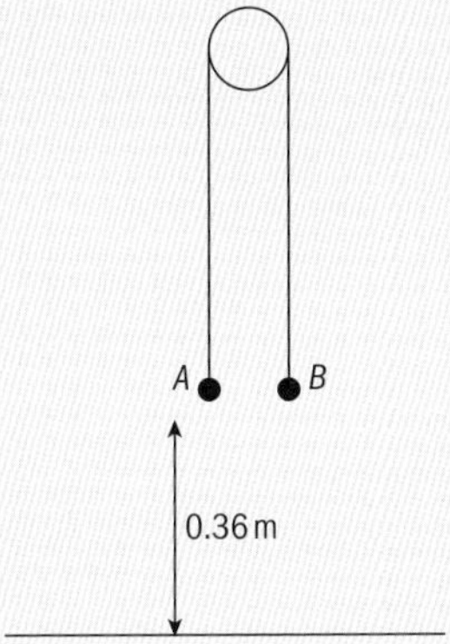

Particles A and B are attached to the ends of a light inextensible string which passes over a smooth pulley. The system is held at rest with the string taut and its straight parts vertical. Both particles are at a height of 0.36 m above the floor (see diagram). The system is released and A begins to fall, reaching the floor after 0.6 s.

i) Find the acceleration of A as it falls. [2]

The mass of A is 0.45 kg. Find

ii) the tension in the string while A is falling, [2]

iii) the mass of B, [3]

iv) the maximum height above the floor reached by B. [3]

Cambridge International AS and A Level Mathematics 9709, Paper 4 Q6 May/June 2009

12. 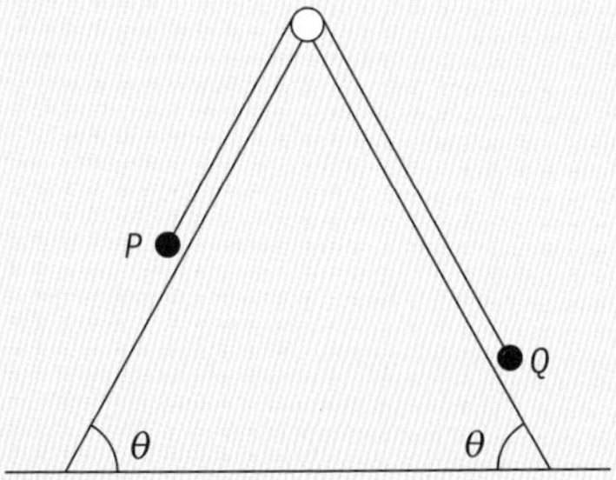

Particles P and Q, of masses 0.6 kg and 0.4 kg respectively, are attached to the ends of a light inextensible string. The string passes over a small smooth pulley which is fixed at the top of a vertical cross-section of a triangular prism. The base of the prism is fixed on horizontal ground and each of the sloping sides is smooth. Each sloping side makes an angle θ with the ground, where $\sin\theta = 0.8$. Initially the particles are held at rest on the sloping sides, with the string taut (see diagram). The particles are released and move along lines of greatest slope.

i) Find the tension in the string and the acceleration of the particles while both are moving. [5]

The speed of P when it reaches the ground is $2\,\text{ms}^{-1}$. On reaching the ground P comes to rest and remains at rest. Q continues to move up the slope but does not reach the pulley.

ii) Find the time taken from the instant that the particles are released until Q reaches its greatest height above the ground. [4]

Cambridge International AS and A Level Mathematics 9709, Paper 41 Q6 May/June 2012

13. A cyclist exerts a constant driving force of magnitude F N while moving up a straight hill inclined at an angle α to the horizontal, where $\sin\alpha = \frac{36}{325}$. A constant resistance to motion of 32 N acts on the cyclist. The total weight of the cyclist and his bicycle is 780 N. The cyclist's acceleration is $-0.2\,\text{ms}^{-2}$.

i) Find the value of F. [4]

The cyclist's speed is $7\,\text{ms}^{-1}$ at the bottom of the hill.

ii) Find how far up the hill the cyclist travels before coming to rest. [2]

Cambridge International AS and A Level Mathematics 9709, Paper 41 Q3 October/November 2013

14.

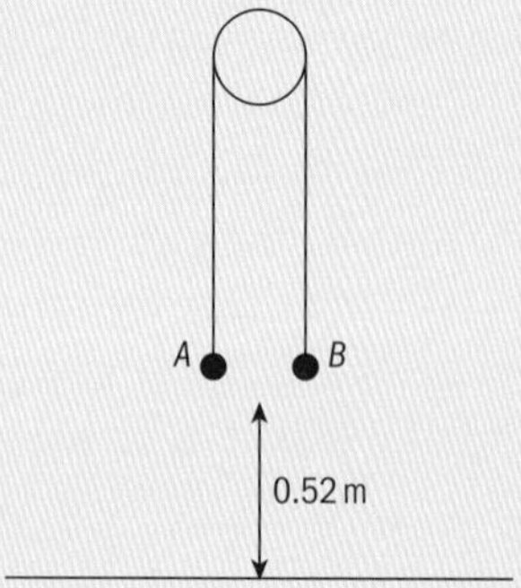

Particles A and B, of masses 0.3 kg and 0.7 kg respectively, are attached to the ends of a light inextensible string. The string passes over a fixed smooth pulley. A is held at rest and B hangs freely, with both straight parts of the string vertical and both particles at a height of 0.52 m above the floor (see diagram). A is released and both particles start to move.

i) Find the tension in the string. [4]

When both particles are moving with speed 1.6 ms^{-1} the string breaks.

ii) Find the time taken, from the instant that the string breaks, for A to reach the floor. [5]

Cambridge International AS and A Level Mathematics 9709, Paper 41 Q6 October/November 2013

15.

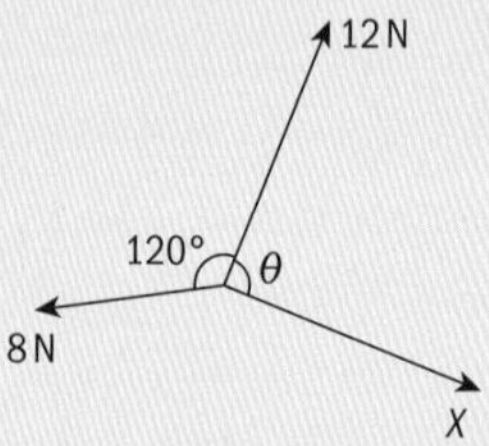

For the following system of forces in equilibrium, find the magnitude of the missing force and size of the angle θ.

16.

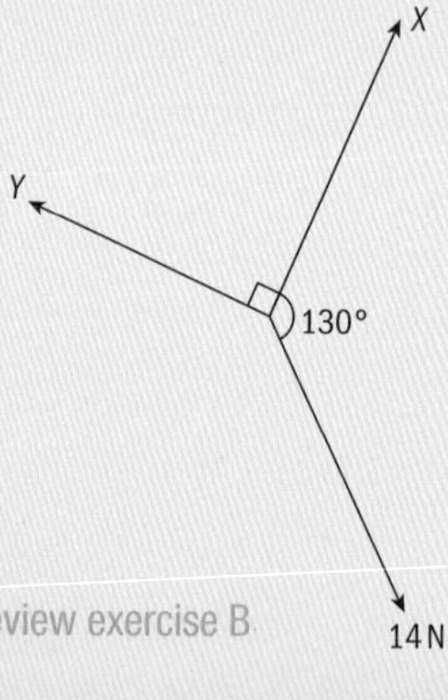

For the following system of forces in equilibrium, find the magnitude of the missing forces.

17.

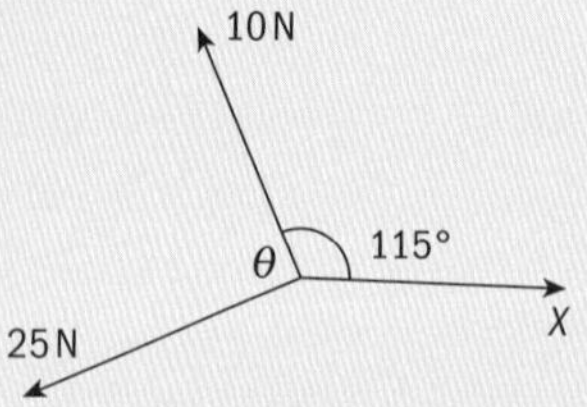

For the following system of forces in equilibrium, find the magnitude of the missing force and the angle θ.

18.

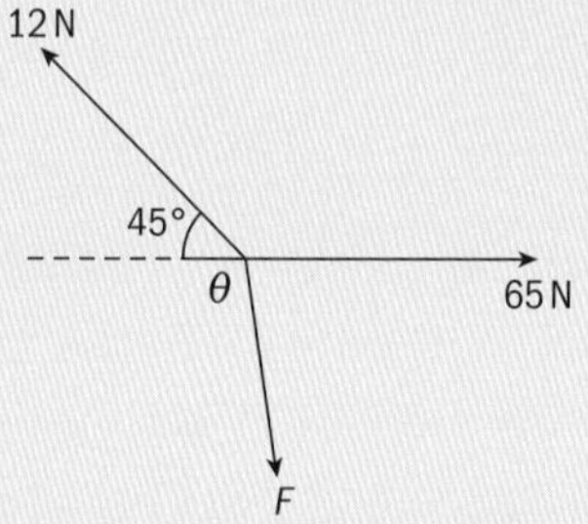

For the following system of forces in equilibrium, find the magnitude of the missing force and the angle θ.

19. The following forces are in equilibrium. Find the missing values which are given as letters.

$$(2\mathbf{i} + 3\mathbf{j})\text{N}, (-5\mathbf{i} + 2\mathbf{j})\text{N}, (a\mathbf{i} + b\mathbf{j})\text{N}$$

20.

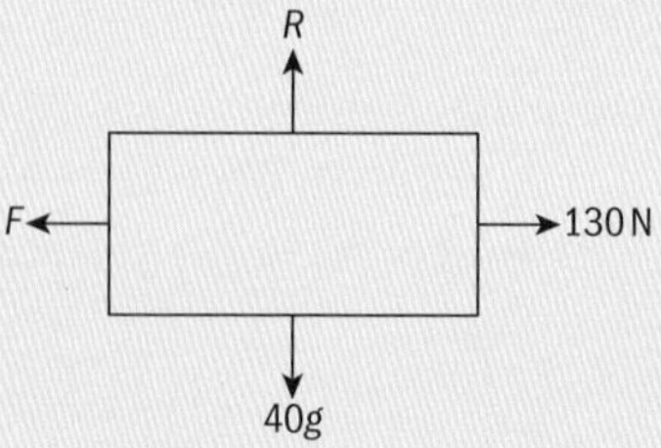

The diagram shows a solid block of wood of mass 40 kg at rest. Given that the coefficient of friction between the block and the surface is 0.35, state whether the forces applied are enough to cause it to move.

21.

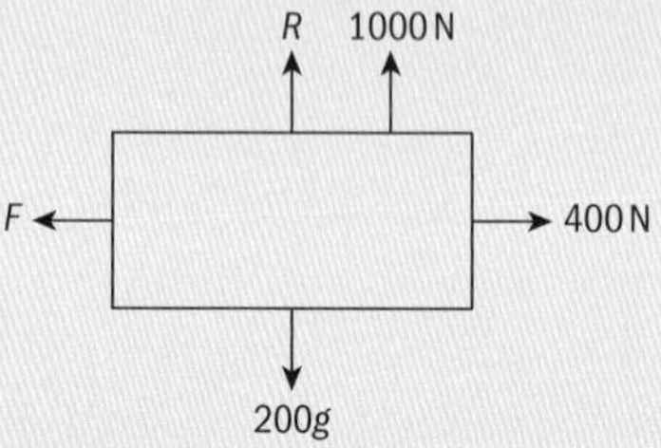

The diagram shows a solid block of rubber of mass 200 kg at rest. Given that the coefficient of friction between the block and the surface is 0.35, state whether the forces applied are enough to cause it to move.

22.

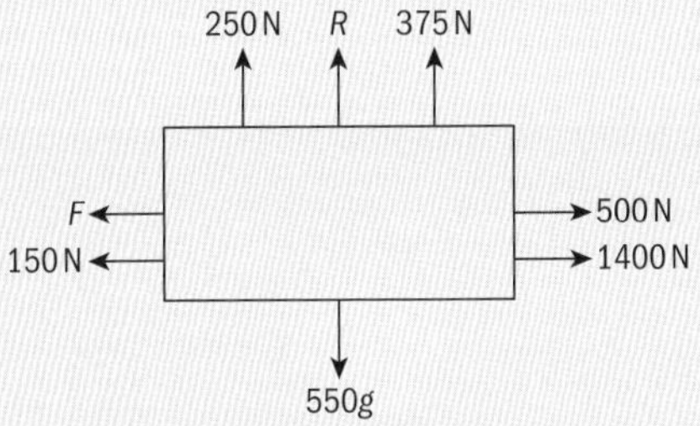

The diagram shows a small car that has been crushed and made into a block which is at rest it is of mass 550 kg. Given that the coefficient of friction between the block and the surface is 0.35, state whether the forces applied are enough to cause it to move.

23.

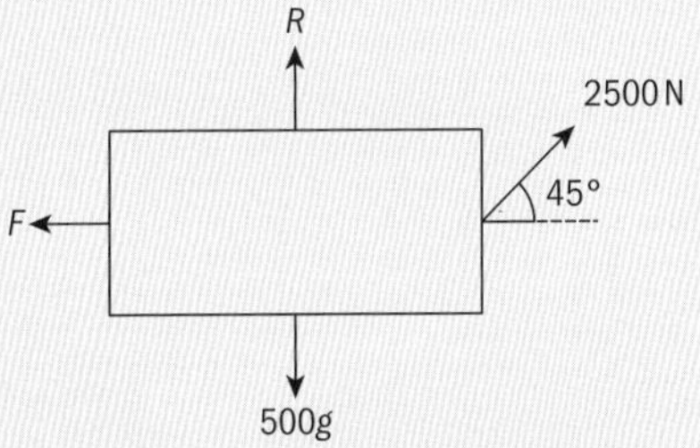

The diagram shows a block of old crushed mobile phones of mass 500 kg at rest. Given that the coefficient of friction between the block and the surface is 0.55, state whether the forces applied are enough to cause it to move.

24.

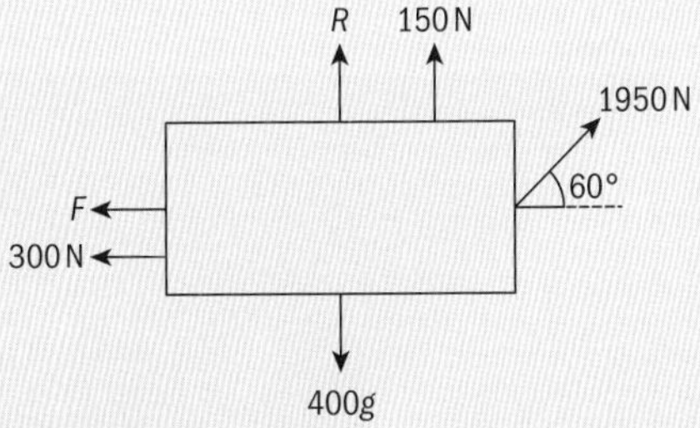

The diagram shows a box of stationary of mass 400 kg at rest. Given that the coefficient of friction between the block and the surface is 0.55, state whether the forces applied are enough to cause it to move.

25.

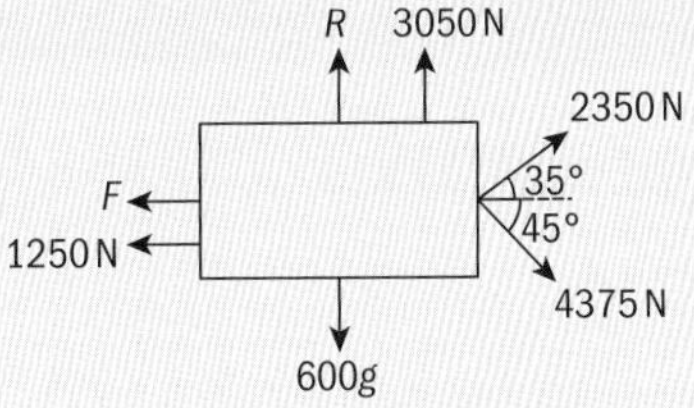

The diagram shows a container of parts for a tractor, of mass 600 kg at rest. Given that the coefficient of friction between the block and the surface is 0.55, state whether the forces applied are enough to cause it to move.

26. A wooden box of max 25 kg rests in limiting equilibrium on a rough inclined slope, 15° to the horizontal. Find the coefficient of friction between the box and the slope.

27. A wooden crate of fish is at rest on a slope which is inclined at 35° to the horizontal. The coefficient between the crate and the slope $\frac{\sqrt{2}}{6}$. Given that the crate is of mass 250 kg, what is its acceleration if a force of 2000 N is applied up the hill along a line of greatest slope?

28.

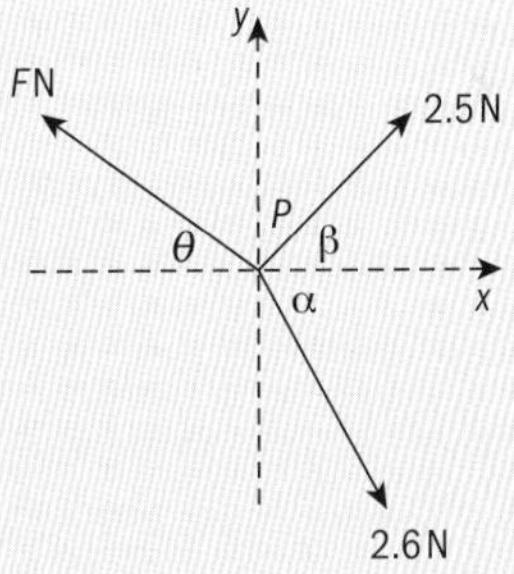

A particle of mass 0.5 kg lies on a smooth horizontal plane. Horizontal forces of magnitudes F N, 2.5 N and 2.6 N act on P. The directions of the forces are as shown in the diagram, where $\tan\alpha = \frac{12}{5}$ and $\tan\beta = \frac{7}{24}$.

i) Given that P is in equilibrium, find the values of F and $\tan\theta$.

ii) The force of magnitude F N is removed. Find the magnitude and direction of the acceleration with which P starts to move.

Cambridge International AS and A Level Mathematics 9709, Paper 41 Q6 May/June 2013

Maths in real-life

Celestial mathematics

Universal theories are a constant challenge to mathematicians and physicists.

People used to believe that the earth was the centre of the Universe. In the 16th and 17th centuries some mathematicians and astronomers developed theories showing that the Earth rotated around the Sun and not vice versa. Copernicus' theories were the first to suggest this.

Whilst Copernicus' theories were essentially correct, they were not entirely accurate. This may be due to the difficulty of collecting observational data when compared with the modern day. Since the 16th century, with the development of technology, there has been significant progression in the development of theories.

Johannes Kepler published his laws of planetary motion at the very end of the 16th century. However the forces which dictate the form of the orbits (gravity) were not identified until almost a century later by Sir Isaac Newton.

The theory of planetary motion also governs the behaviour of objects like satellites. These are essential for our modern communications systems to work. Satellites which orbit at about 23 000 miles above the Earth complete one revolution around the Earth in 24 hours. This means that they can stay precisely above a particular location on Earth as it spins on its axis. This is called a geostationary orbit and allows a satellite receiver simply to be pointed in one direction to always receive the transmission from that satellite.

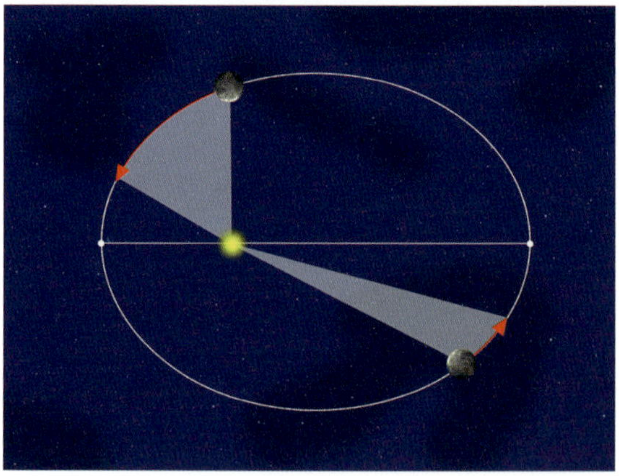

▲ Kepler proposed that the radius vector from Sun to planet sweeps out equal areas in equal time during the elliptical orbit. One consequence of this is that the Earth travels faster when it is close to the Sun than when it is further away.

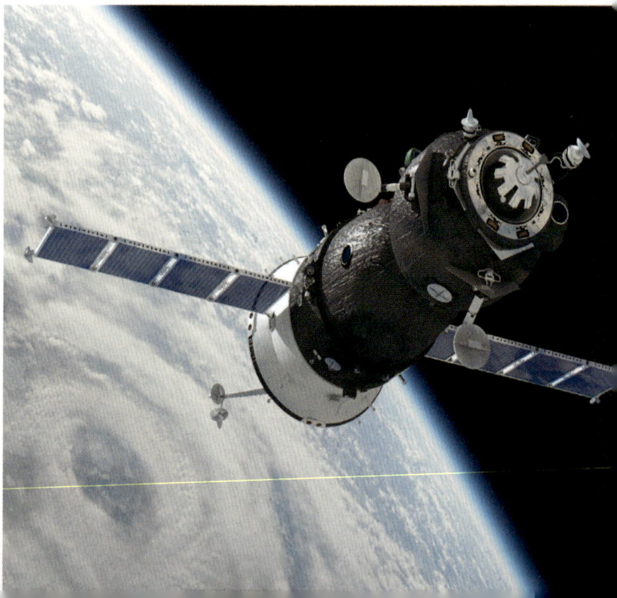

Newton's theory of gravitation was the basis for all cosmological models for over 200 years until Einstein's proposed his theory of relativity in 1905. This picture illustrates one of the key results in Einstein's theory – that light will bend round massive objects like stars.

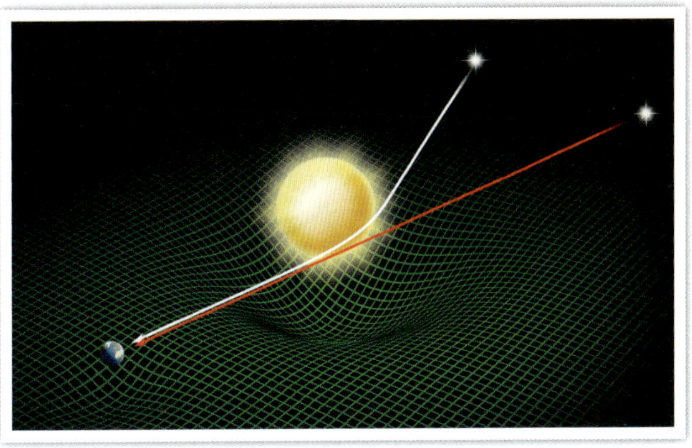

Scientists and mathematicians continue to work to develop more complete theories of how the universe and its component parts work together. The discovery of the Higgs boson in 2012 caused unheard of interest among the public in a scientific advance.

The Higgs boson earned Francois Englert and Peter Higgs a Noble Prize. It was theorised in 1964, but took 40 years to be discovered. It is a very small particle that explains why things have mass and why they hold together. In succession this explains why we are able to exist. This particle is part of 'The Standard Model'of particle physics. This set of rules lays out our understanding of the fundamental basis of the universe.

The search for the Higgs boson began in the United States in the 1980s. They attempted to build a giant atom collider beneath the Texas prairie. However, the project became a logistical nightmare, costing $2 billion, and was stopped by Congress in 1993.

▼ Upon its discovery, it was thought that the Higgs boson could lead to deeper understanding of the Universe. It was also believed that the technology that was used to discover it could be applied to other areas of research.

7 Work and energy

The Japanese bullet train travels at very high speeds, up to 300 km/h. The train's engine does **work** so energy is transferred to create the motion. The bullet train is so-called because of its shape and speed. The engineers who designed this train would have made use of the kind of formulae we are learning about to calculate how much work the engine would need to do in order for the train to achieve its high speeds.

Objectives

- Understand the concept of the work done by a force, and calculate the work done by a constant force when its point of application undergoes a displacement not necessarily parallel to the force.
- Understand the concepts of gravitational potential energy and kinetic energy, and use appropriate formulae.
- Understand and use the relationship between the change in energy of a system and the work done by external forces, and use in appropriate cases the principle of conservation of energy.

Before you start

You should know how to:

1. Resolve a force in a given direction.

e.g. Find the resolved part of the force in the direction of the dotted arrow.

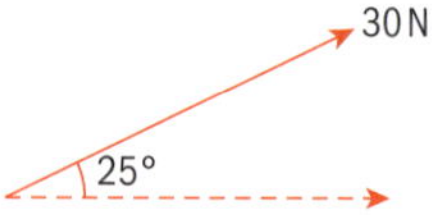

The resolved part is $30\cos 25°$
$= 27.18923361 = 27.2\text{ N}$ (to 3 s.f.)

2. Use Newton's first law.

e.g. A parachutist of mass 85 kg is falling with constant speed. Find the magnitude of the resistive force experienced by the parachutist.

Because of Newton's first law, the resultant force on the parachutist is 0.
Hence, if R is the resistive force, then
$R - 85g = 0$
Hence $R = 850\text{ N}$ (taking g as 10 ms^{-2}).

Skills check:

1. Find the resolved part of each force in the direction of the dotted arrow.

a)

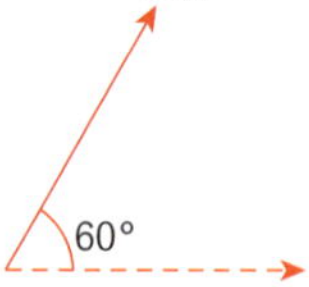

b)

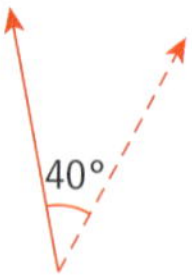

c)

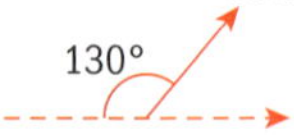

2. a) A diver of mass 73 kg is rising from the seabed at a constant speed. Find the upward force acting on the diver.

b) A lorry is ascending a hill inclined at 2° to the horizontal at a constant speed. The mass of the lorry is 12 000 kg. The resistance to motion is constant and has magnitude 400 N. Find the driving force provided by the lorry's engine.

7.1 Work

When a force moves the point to which it is applied, it is said to do **work**. Consider a constant force F, acting on a particle of mass m. Suppose the particle has an initial speed $u\,\text{ms}^{-1}$ and a final speed of $v\,\text{ms}^{-1}$, with the force acting over a distance s metres.

Then by Newton's second law, $F = ma$

Since a is a constant then, $v^2 = u^2 + 2as$

Rearranging gives $as = \frac{1}{2}v^2 - \frac{1}{2}u^2$

Multiplying by m, $mas = \frac{1}{2}mv^2 - \frac{1}{2}mu^2$

Now, $Fs = mas$

Hence, $Fs = \frac{1}{2}mv^2 - \frac{1}{2}mu^2$ (1)

If you were to lift an object, push a heavy box or pedal a bicycle then you would know in each case that you were doing **work**. Work is done whenever a force is applied to an object to change its motion or its position.

The product of F and s is referred to as the **work done** by the constant force F as the particle moves through a displacement s.

The S.I. unit of work is the **joule (J)**, which is the amount of work done when a force of 1 newton moves a distance of 1 metre. The joule is named after the English physicist **James Prescott Joule (1818 – 89)**.

Did you know?
S.I. is an abbreviation for *Le Système international d'unités* – the International System of Units.

Example 1

An object of mass 40 kg is pulled a distance of 5 m at constant speed across horizontal rough ground by means of a horizontal rope. The coefficient of friction between the object and the ground is 0.3. Find the work done by each of the forces acting on the object.

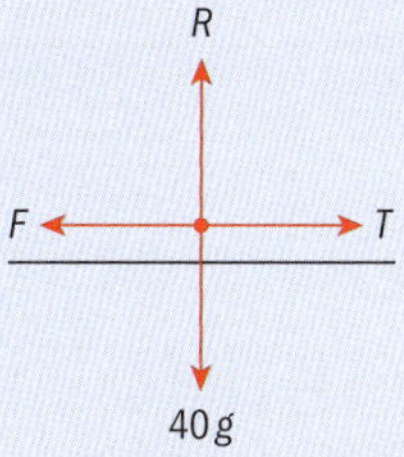

'By means of' is the same as saying 'with the use of'.

Continued on the next page

Resolving vertically, we obtain $R - 40g = 0$

$R = 400\,\text{N}$

Using $F = \mu R$, where $\mu = 0.3$, then $F = 120\,\text{N}$

Resolving horizontally, we obtain $T - F = 0$

$T = F = 120\,\text{N}$

There is no vertical displacement of the object, so the work done by the weight (400 N) and by R is zero.

The displacement in the direction of T is 5 m, so the work done by the tension $= 120 \times 5 = 600\,\text{J}$

The displacement in the direction of F is -5 m, so the work done by the friction $= 120\ -5 = -600\,\text{J}$

Notice the negative work done by friction. We can say the work done by friction is -600 J, or usually that there is 600 J of work done *against* friction.

It is important to realize that the displacement must take place in the direction of the force. Hence, any forces that are perpendicular to the displacement do no work.

Frequently, the applied force is directed at an angle to the direction that the displacement occurs. For the object in Example 1, the rope used to pull the object may be inclined to the horizontal.

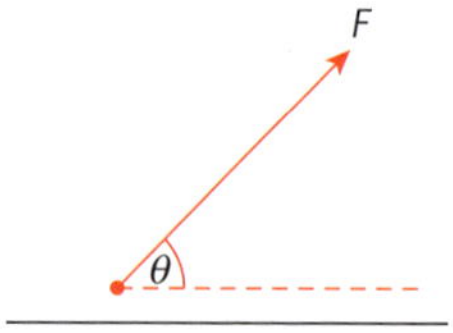

Suppose a force F is applied to an object that is then displaced by a distance s in a direction making an angle θ with the direction of F.

The force F can be resolved into two components: parallel and perpendicular to the direction of the displacement. The perpendicular component, $F\sin\theta$, does no work because there is no displacement in that direction. The parallel component, $F\cos\theta$, is displaced a distance s. Therefore the work done by F is

$(F\cos\theta) \times s = Fs\cos\theta$

Example 2

A packing case is pulled along horizontal ground a distance of 6 m by means of a rope inclined at 35° to the horizontal. The tension in the rope is 400 N. Find the work done by the tension.

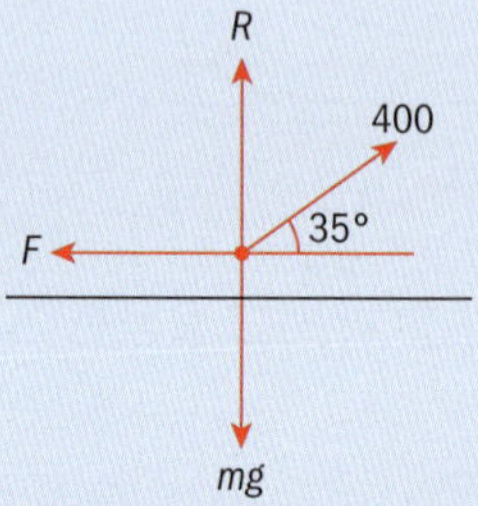

The component of the force in the direction of the motion is $400\cos 35°$.
The displacement in this direction is 6 m.
Hence, work done $= 400\cos 35° \times 6 = 1965.964906 = 1970\,\text{J}$ (to 3 s.f.)

Exercise 7.1

1. A body is pulled a distance of 20 m across a horizontal surface against a resistance of 40 N. If the body moves with constant velocity, find the work done against the resistance.

2. Find the work done by a crane lifting a load of 200 kg at constant speed through a distance of 6.4 m.

3. Find the work done in pulling a packing case of mass 70 kg a distance of 18 m against a resistance of 150 N on a horizontal surface.

4. A force of 80 N moves an object 15.8 m in the direction of the force. Find the work done by the force.

5. Find the work done by a man of mass 78 kg in climbing vertically up a ladder 5 m high at constant speed.

6. A force of 50 N acts on a block at an angle of 30° above the horizontal. The block moves a horizontal distance of 3 m. Find the work done by the applied force.

7. Calculate the work done by a 3 N force, directed at an angle 40° to the upward vertical, to move a box 400 cm across a horizontal floor at constant speed.

8. A woman pushes a package 20 m along level ground at constant speed. The work done by the woman is 160 J. Find the average force resisting the motion.

9. A car is towed at constant speed by means of a tow rope angled at 10° to the horizontal. The work done by the tension in the rope in moving the car 100 m is 2000 J. Find the tension in the rope.

10. A body of mass 8 kg is at rest on a rough horizontal surface. The coefficient of friction between the body and the surface is 0.4. A force of 120 N is applied for a period of 10 s in each of the following cases:

    a) 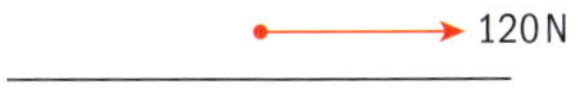

    b) 

    In each case, find the distance moved by the body and the work done by the 120 N force.

7.2 Kinetic energy

If we refer back to equation (1),

$$Fs = \frac{1}{2}mv^2 - \frac{1}{2}mu^2$$

the left hand side was the work done by the force F. The right hand side of the equation has two terms of a similar form.

If an object is moving with a speed v, the quantity $\frac{1}{2}mv^2$ is called the **kinetic energy** of the body.

So the kinetic energy of a body is the energy it possesses because of its motion. Equation (1) tells us that when a force does work on a body so as to increase its speed, then the work done is a measure of the increase in the kinetic energy of the body.

Example 3

A car of mass 1600 kg is travelling along a straight horizontal road at 15 ms^{-1}. The brakes are applied as the car approaches a junction. The car travels 25 m before coming to rest. Find

a) the initial kinetic energy of the car

b) the work done in stopping the car

c) the average force applied in stopping the car.

a) The initial kinetic energy of the car $= \frac{1}{2} \times 1600 \times (15)^2 = 180\,000$ J

b) The work done in stopping the car = change in kinetic energy = 180 000 J

c) Since work done = force × displacement, then average force $= \frac{180000}{25} = 7200$ J

Work done and kinetic energy are both measured in joules (J).

Example 4

A body of mass 6 kg increases its energy by 38 J. If its initial speed was 2 ms^{-1}, find its final speed.

Initial kinetic energy $= \frac{1}{2}mv^2$

$= \frac{1}{2} \times 6 \times (2)^2 = 12$ J

Final kinetic energy = 12 + 38 = 50 J

If the final speed of the body is v, then $50 = \frac{1}{2} \times 6 \times v^2$

$v = 4.082482905 = 4.08$ ms^{-1} (to 3 s.f.)

Exercise 7.2

1. Find the kinetic energy of
   a) a body of mass 5 kg moving with speed 4 ms^{-1}
   b) a body of mass 2 kg moving with speed 3 ms^{-1}
   c) a car of mass 1200 kg moving with speed 10 ms^{-1}
   d) a particle of mass 100 g moving with speed 20 ms^{-1}.

2. Find the gain in kinetic energy when
   a) a car of mass 1.4 tonnes increases its speed from 5 ms^{-1} to 6 ms^{-1}
   b) a body of mass 5 g increases its speed from 200 ms^{-1} to 300 ms^{-1}.

3. Find the loss in kinetic energy of
   a) an object of mass 5 kg which decreases its speed from 3 ms^{-1} to 2 ms^{-1}
   b) a car of mass 900 kg which decreases its speed from 14 ms^{-1} to 10 ms^{-1}.

4. A body of mass 12 kg is moving with speed 4 ms^{-1}. The body's kinetic energy increases by 80 J. Find the final speed of the body.

5. When a squash ball bounces off a wall it loses 20% of its kinetic energy. A squash ball has a mass of 25 g and hits the wall at 4 ms^{-1}. Find
   a) the kinetic energy of the ball immediately before it hits the wall
   b) the speed of the ball immediately after it rebounds from the wall.

6. A man pushes a car of mass 1200 kg from rest. He exerts a horizontal force of 150 N. Find
   a) the work done on the car as it moves 10 m
   b) the speed of the car after 10 m.

'Exert' means 'apply'.

7.3 Gravitational potential energy

When a climber scales a rock face, they are doing work against the force of gravity. The climber is said to be increasing their **gravitational potential energy**.

The word 'scale' has many mathematical meanings – but here, it means 'climb'.

Suppose we lift an object of mass 15 kg through a height of 8 m. The work we do against gravity is $15g \times 8 = 1200$ J.

If we now allow the object to fall back to its original level, its weight will do 1200 J of work.

Generalizing this to the case of a mass m kg, raised vertically through a vertical height h m, the work done against gravity is mass $\times\ g \times$ height.

So mgh is the work done against gravity.

If you climb the stairs, then the work done is your weight multiplied by the vertical height gained. In fact the path taken between two points is irrelevant in calculating the work done. The work done is always $mg \times$ (vertical height gained), and is independent of the path taken to gain the height.

By lifting the object we gave the object energy. The energy given to it, which depends on the position of the object in a gravitational field, is called **gravitational potential energy (GPE)**.

Raising an object increases its GPE, while lowering it decreases its GPE. The amount of energy is simply mgh, where h is the distance above some arbitrary reference point. When an object loses height, this potential energy is converted to kinetic energy.

Example 5

A carriage on a roller coaster ride has a mass of 130 kg. Find the change in potential energy when the carriage descends 12 m.

Change in potential energy $= mgh = 130g \times 12 = 15\,600$ J

Gravitational potential energy is also measured in joules (J).

Example 6

A climber of mass 75 kg scales a mountain 1.8 km high. Find her gain in potential energy.

Increase in potential energy $= mgh = 75g \times 1800 = 1\,350\,000$ J $= 1350$ kJ

1 kJ = 1000 J, and is called a **kilojoule**.

7.4 Conservation of energy

If, as an object moves, all the potential energy is converted to kinetic energy, then the situation is called a **conservative system**. If some work is done, so that not all the potential energy is converted to kinetic energy, this is a **non-conservative system**.

So the total energy of a system remains constant provided no external work is done and there are no sudden changes in the motion of the system.

Example 7

A ball of mass 0.6 kg is dropped from rest at a height of 10 m above the ground. Neglecting air resistance, find

a) the loss in gravitational potential energy in falling to the ground

b) the gain in kinetic energy at the instant the ball reaches the ground

c) the speed with which the ball hits the ground.

a) The loss in GPE = $mgh = 0.6\,g \times 10 = 60\,\text{J}$

b) Because of conservation of energy, the kinetic energy gained is equal to the GPE lost. So the gain in kinetic energy is 60 J.

c) The change in kinetic energy is $\frac{1}{2}mv^2 - \frac{1}{2}mu^2$

$$= \frac{1}{2} \times 0.6 \times v^2 - \frac{1}{2} \times 0.6 \times (0)^2 = 60$$

$$v = 14.14213562 = 14.1\,\text{ms}^{-1} \text{ (to 3 s.f.)}$$

If an external force acts on a system so that work is done, we can still make use of energy to solve the problem because the total work done on the system equals the change in energy.

Exercise 7.4

1. A base jumper with mass 85 kg jumps from the top of a building, 300 m above the ground. He falls with an initial velocity of $4\,\text{ms}^{-1}$ towards the ground. He releases his parachute at a point 80 m above the ground. Find

a) the initial kinetic energy of the jumper

b) the potential energy lost by the jumper in moving from the top of the building to the point where he releases his parachute

c) his speed at the instant that he releases his parachute, stating an assumption you have made in modelling this situation.

Did you know?

Base jumping is a sport in which participants jump from a fixed object (such as a building or a cliff) and use a parachute to break their fall.

2. A particle of mass 0.4 kg is projected up a smooth plane inclined at an angle θ to the horizontal, where $\tan\theta = \frac{5}{12}$. The particle moves through a point A at a speed of $12\,\text{ms}^{-1}$. The particle continues to move up the line of greatest slope and comes to instantaneous rest at a point B.

Find

a) the height of B above the level of A

b) the distance AB

c) the speed of the particle when it returns to A.

3. A parachutist of mass 70 kg jumps from a height of 1000 m and hits the ground at 6 ms^{-1}.

Find

a) the potential energy lost

b) the work done against the resistive forces during the jump.

4. A body of mass 2 kg is released from rest and falls freely under gravity. Find its speed when it has fallen a distance of 15 m.

5. A body of mass 5 kg is released from rest and falls freely under gravity. Find the distance it has fallen when its speed is 8 ms^{-1}.

6. A stone of mass 0.2 kg is dropped down a well. The stone hits the surface of the water with a speed of 16 ms^{-1}.

   a) Calculate the kinetic energy of the stone as it hits the water.

   b) Find the height above the water from which the stone was dropped.

   c) When the stone hits the water, it begins to sink vertically and experiences a constant resistance of 18 N. Find the depth the stone has sunk to when the speed of the stone is 4 ms^{-1}.

7. A body of mass 4 kg is projected vertically downwards with speed 2 ms^{-1}. Find the speed of the body as it passes through a point 6 m below the point of projection. Assume there are no resistances to the motion.

8. A and B are two points in a vertical line with A above B. A body of mass 0.6 kg is released from A and falls vertically, passing through B with speed 12 ms^{-1}. Find the distance AB. Assume there are no resistances to the motion.

9. A and B are two points in a vertical line with A above B. A body of mass 0.5 kg falls vertically. It passes through A with speed 2 ms^{-1} and passes through B with speed 8 ms^{-1}. Find the distance AB. Assume there are no resistances to the motion.

10. A and B are two points in a vertical line with A, a distance of 4 m above B. A particle P, of mass m kg, is projected vertically upwards from B with a speed of 15 ms^{-1}. Assuming there are no resistances to motion, find

  a) the speed of P as it passes through A

  b) the speed of P as it passes through A again, travelling downwards

  c) the speed of P as it passes through B

  d) the height above B of the highest point reached.

11. A smooth slope is inclined at $\tan^{-1}\left(\frac{3}{4}\right)$ to the horizontal. A particle of mass 0.4 kg is released from rest at the top of the slope. The particle reaches the bottom of the slope with speed 8 ms^{-1}. Find the length of the slope.

12. Point A is at the bottom of a smooth slope which is inclined at an angle θ to the horizontal, where $\tan\theta = \frac{7}{24}$. A particle is projected from A with speed 16 ms^{-1} up the line of greatest slope of the plane and passes through a point B with speed 3 ms^{-1}. Find the distance AB.

7.5 The work–energy principle

As an object moves, if some work is done but not all the potential energy is converted to kinetic energy, this is a **non-conservative system**. In these situations we can still make use of energy to solve a problem because the total work done on the system is equal to the change in energy.

Example 8

A particle of mass 4 kg is projected down a plane inclined at 30° to the horizontal with speed 1 ms^{-1}. There is a constant resistance of 5 N. Find the speed of the particle after it has travelled 6 m down the plane.

The vertical height lost by the particle is $6\sin 30° = 3$ m

In moving down the slope the loss in PE of the particle is $mgh = 4g \times 3 = 120$ J

The gain in KE is $\frac{1}{2}mv^2 - \frac{1}{2}mu^2 = \frac{1}{2} \times 4 \times v^2 - \frac{1}{2} \times 4 \times (1)^2 = 2v^2 - 2$

The work done against the resistance force is $Fs = 5 \times 6 = 30$ J

Hence, by conservation of energy, the GPE lost is converted into KE and is used to overcome the resistance

$$120 = 2v^2 - 2 + 30$$

$$v = 6.782329983 = 6.78 \text{ ms}^{-1} \text{ (to 3 s.f.)}$$

Example 9

A car of mass 1200 kg has a speed of 26 ms^{-1} at the bottom of a hill inclined at 5° to the horizontal. The car travels up a line of greatest slope of the hill. After a distance of 500 m the car's speed has decreased to 12 ms^{-1}. The resistance to motion is constant and has magnitude 400 N. Find the constant driving force produced by the car's engine.

The vertical height gained by the car is 500 sin 5° = 43.6 m (to 3 s.f.)

In moving up the hill, the gain in PE of the car is $mgh = 1200 \times g \times 43.6 = 522\,934.5$ J

The gain in KE is $\frac{1}{2}mv^2 - \frac{1}{2}mu^2 = \frac{1}{2} \times 1200 \times (12)^2 - \frac{1}{2} \times (26)^2$

$$= \frac{1}{2} \times 1200 \times 144 - \frac{1}{2} \times 1200 \times 676$$
$$= 86\,400 - 405\,600$$
$$= -319\,200 \text{ J}$$

As this value is negative, KE is lost as the car travels up the hill, and the *change* in KE is 319 200 J.

The work done against the resistance force of 400 N is $(D - 400) \times s = (D - 400) \times 500$

By conservation of energy, the KE lost is converted into PE and used to overcome the resistance.

So, $522\,934.5 = 319\,200 + (D - 400) \times 500$

$$(D - 400) \times 500 = 203\,634.5$$
$$D - 400 = 407\,269$$
$$D = 807.269$$

The driving force, D, is 807 J (to 3 s.f.)

There are cases where the energy done by a specific force is given. The following example shows how to deal with this type of problem. It cannot be assumed that the force is constant and hence the use of constant acceleration formulae is prohibited.

Example 10

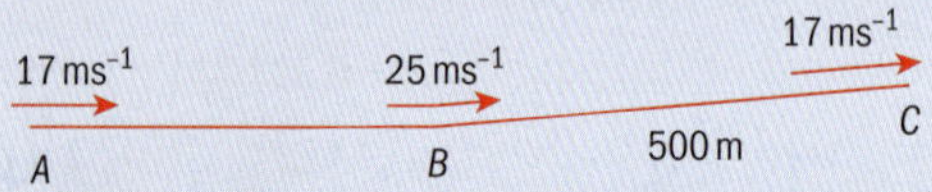

A lorry of mass 12 500 kg travels along a road that has a straight horizontal section AB and a straight inclined section BC. The length of BC is 500 m. The speeds of the lorry at A, B and C are 17 ms^{-1}, 25 ms^{-1} and 17 ms^{-1} respectively (see diagram).

i) The work done against the resistance to motion of the lorry, as it travels from A to B, is 5000 kJ. Find the work done by the driving force as the lorry travels from A to B.

ii) As the lorry travels from B to C, the resistance to motion is 4800 N and the work done by the driving force is 3300 kJ. Find the height of C above the level of AB.

Cambridge AS and A Level Mathematics 9709
Paper 4 Q5 May/June 2007

▶ Continued on the next page

The temptation in this question is to perhaps use constant acceleration equations. However this would require the assumption that the driving force and the resistance are constant, which is not stated in the question. Only the work done by the resistance and driving force is given. This is something to be aware of in questions of this type.

i) The gain in KE is

$\frac{1}{2}mv^2 - \frac{1}{2}mu^2 = \frac{1}{2} \times 12\,500 \times (25)^2 - \frac{1}{2} \times 12\,500 \times (17)^2 = 2\,100\,000\text{ J}$

Work done by Driving Force = Gain in KE + work done against resistance

$= 2\,100\,000 + 5\,000\,000 = 7\,100\,000\text{ J} = 7100\text{ kJ}$

ii) The loss in KE =

$\frac{1}{2}mv^2 - \frac{1}{2}mu^2 = \frac{1}{2} \times 12\,500 \times (25)^2 - \frac{1}{2} \times 12\,500 \times (17)^2 = 2\,100\,000\text{ J}$

The work done against resistance = $4800 \times 500 = 2\,400\,000\text{ J}$

The gain in PE = $12\,500 \times g \times h = 125\,000h$

Work done by driving force + loss in KE = gain in PE + work done against resistance

$3\,300\,000 + 2\,100\,000 = 125\,000h + 2\,400\,000$

$h = 24\text{ m}$.

Exercise 7.5

1. A and B are two points 8 m apart on a horizontal smooth surface. A particle of mass 2 kg is initially at rest at A and is pushed by a force of constant magnitude acting in the direction from A to B. The particle reaches a point B with speed 6 ms^{-1}. Find the magnitude of the force.

2. A car of mass 1000 kg descends a hill angled at θ to the horizontal, where $\sin\theta = 0.1$. There is a constant resistance to motion of magnitude 200 N. Find the work done by the brakes in bringing the car to rest from a speed of 9 ms^{-1} in a distance of 50 m. Assuming that the braking force is constant, find the braking force produced by the car.

3. In a downhill ski race, competitors descend from a start point, which is 1800 m above sea level, to a finish line, which is 1100 m above sea level. A competitor has a total mass of 85 kg including equipment. He starts from rest and crosses the finish line with a speed of 12 ms^{-1}. Find the work done in overcoming the resistances to his motion.

4. A lorry of mass 5000 kg accelerates from 5 ms^{-1} to 8 ms^{-1} while covering a distance of 60 m on a horizontal road. The resistance to motion is constant and of magnitude 250 N. Find the driving force.

5. Find the force needed to accelerate a train of mass 400 tonnes from 12 ms^{-1} to 20 ms^{-1} in a distance of 2 km along a horizontal track, assuming the resistance to motion is constant and 150 000 N.

6. A gymnast of mass 60 kg swings on a rope of length 12 m. Initially the rope makes an angle of 50° with the vertical. Find
   a) the decrease in his potential energy when the rope is vertical
   b) the speed of the gymnast when the rope is vertical.

7. *A* and *B* are two points 6 m apart on a horizontal smooth surface. A particle of mass 5 kg is initially at rest at *A* and is pushed towards *B* by a constant force of magnitude 12 N. Find the speed of the particle at B.

8. A child of mass 28 kg goes down a slide, starting from rest. The total drop in height of the slide is 4.5 m.
   a) If the slide is smooth, find the speed of the child at the bottom of the slide.
   b) In fact the slide is rough and the child reaches the bottom travelling with speed 5 ms^{-1}. Find the work done against friction and the average friction force, given that the total length of the slide is 12 m.

9. A constant force of magnitude 15 N pushes a body of 6 kg in a straight line across a smooth horizontal surface. The body passes through a point *A* with speed 3 ms^{-1} and then through a point *B*, 4 m from *A*. For the motion of the body from *A* to *B*, find
   a) the work done by the 15 N force
   b) the final speed of the body.

10. A skateboarder goes down a ramp formed by the arc of a circle of radius 8 m. The total mass of the skateboarder and her board is 65 kg. She starts from rest at *A*, the top of the ramp. Find the speed with which she leaves the ramp at *B*, given that there is a constant resistance of 35 N.

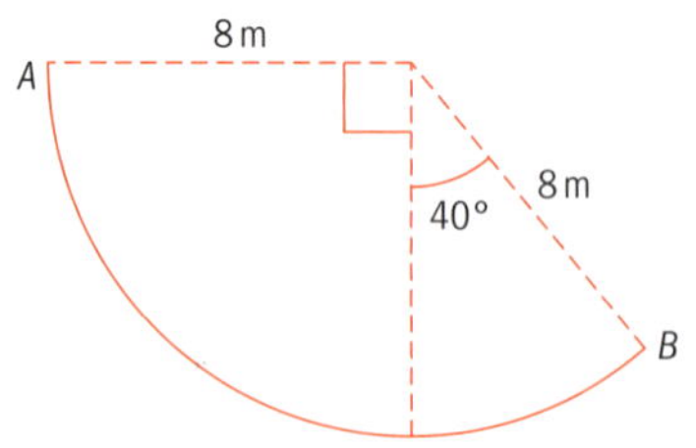

Summary exercise 7

1.

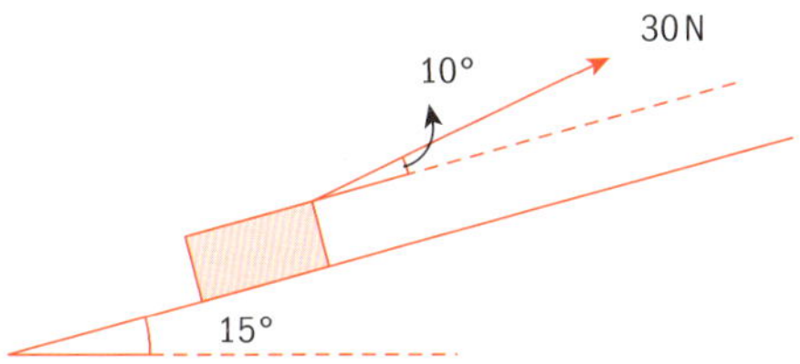

A box of mass 8 kg is pulled, at constant speed, up a straight path which is inclined at an angle of 15° to the horizontal. The pulling force is constant, of magnitude 30 N, and acts upwards at an angle of 10° from the path (see diagram). The box passes through the points *A* and *B*, where *AB* = 20 m and *B* is above the level of *A*. For the motion from *A* to *B*, find

i) the work done by the pulling force, [2]

ii) the gain in potential energy of the box, [2]

iii) the work done against the resistance to motion of the box. [1]

Cambridge International AS and A Level Mathematics 9709, Paper 4 Q1 October/November 2006

2. *A* and *B* are two points on a line of greatest slope of a smooth inclined plane, with *B* a vertical distance of 9 m below the level of *A*. A particle of mass 0.6 kg is projected down the plane from *A* with a speed of 1.5 ms^{-1}. Find the speed of the particle when it reaches *B*.

3. A load of mass 160 kg is lifted vertically by a crane, with constant acceleration. The load starts from rest at the point *O*. After 7 s, it passes through the point A with speed 0.5 ms^{-1}. By considering energy, find the work done by the crane in moving the load from *O* to *A*. [6]

Cambridge International AS and A Level Mathematics 9709, Paper 4 Q4 October/November 2008

EXAM-STYLE QUESTION

4. A box of mass 40 kg is dragged across a horizontal floor by a constant force of magnitude 300 N acting at an angle θ above the horizontal. The total resistance to motion has magnitude 250 N. The box starts from rest at a point *A*, and passes a point *B*, 20 m from *A*, with a speed of 1.5 ms^{-1}.

a) For the box's motion from *A* to *B*, find

i) the increase in the kinetic energy of the box

ii) the work done against the resistance to motion of the box.

b) Hence calculate the value of θ.

5. A car of mass 1250 kg travels from the bottom to the top of a straight hill of length 600 m, which is inclined at 2.5° to the horizontal. The resistance to motion of the car is constant and equal to 400 N. The work done by the driving force is 450 kJ. The speed of the car at the bottom of the hill is 30 ms^{-1}. Find the speed of the car at the top of the hill. [5]

Cambridge International AS and A Level Mathematics 9709, Paper 41 Q2 May/June 2013

6. A skier of mass 70 kg is pulled up a slope, which makes an angle of 15° with the horizontal. The skier is subject to a constant resistance of 50 N. The skier's speed at a point A on the slope is $1\ \text{ms}^{-1}$ and, later, at a point B his speed is $2.5\ \text{ms}^{-1}$. The distance AB is 35 m. Find the work done by the pulling force on the skier as he moves from A to B.

EXAM-STYLE QUESTIONS

7. A girl on a sledge slides down a slope of length 270 m which descends a vertical distance of 60 m. The mass of the girl and sledge in total is 45 kg. The speed of the sledge at the top of the slope is $2\ \text{ms}^{-1}$ and the speed at the bottom is $5.4\ \text{ms}^{-1}$. Given that the resistance to motion is constant, find this resistance.

8. A car of mass 1400 kg travels up the line of greatest slope of a hill inclined at 2° to the horizontal. The car passes through a point A with speed $5\ \text{ms}^{-1}$ and through a point B with a speed of $10\ \text{ms}^{-1}$. Given that the car's engine produces a constant force of 1200 N and the resistance to motion is constant and of magnitude 250 N, find the distance AB.

9. 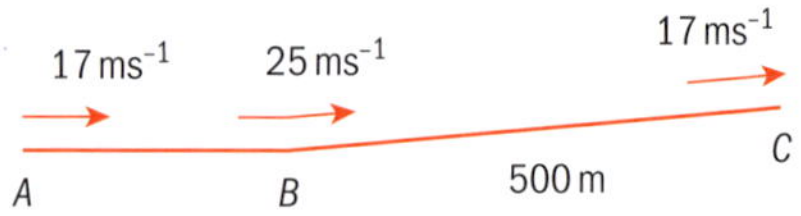

A lorry of mass 12 500 kg travels along a road that has a straight horizontal section AB and a straight inclined section BC. The length of BC is 500 m. The speeds of the lorry at A, B and C are $17\ \text{ms}^{-1}$, $25\ \text{ms}^{-1}$ and $17\ \text{ms}^{-1}$ respectively (see diagram).

i) The work done against the resistance to motion of the lorry, as it travels from A to B, is 5000 kJ. Find the work done by the driving force as the lorry moves from A to B. [4]

ii) As the lorry travels from B to C, the resistance to motion is 4800 N and the work done by the driving force is 3300 kJ. Find the height of C above the level of AB. [4]

Cambridge International AS and A Level Mathematics 9709, Paper 4 Q5 May/June 2007

Chapter summary

- Work done by a force F is Fs, where s is the distance moved in the direction of the force.
- Forces perpendicular to the motion do no work.
- Work done by a force F N acting at an angle θ to the direction of motion is $F\cos\theta s$, where s m is the distance moved in the direction of the force.
- The kinetic energy (KE) of a body of mass m kg moving with speed v ms^{-1} is given by $\frac{1}{2}\text{m}v^2$.
- The gravitational potential energy (GPE) of a body of mass m kg at a height h above a given reference point is mgh.
- Energy is conserved when no forces other than gravity do work.
- The work–energy principle means that the change in energy is equal to the total work done by the forces acting on the body.
- When specific values of work done by a force are given, then the force cannot be assumed to be constant. The work-energy principle is the only method of solving problems of this type.
- The S.I. unit of energy is the joule (J).

8 Power

A Formula One racing car can develop a great deal of **power**. A top speed of over 300 km h^{-1} and the engine generating over 560 kW of power suggests that the power of the engine is related to the speed of the car. As a racing car travels faster and faster, comparatively more power is required because of the increase in air resistance. In order to build these cars, the engineers working on them need to know exactly how fast the car will travel when the engine is producing a certain amount of power.

Objectives

- Use the definition of power as the rate at which a force does work, and use the relationship between power, force and velocity for a force acting in the direction of motion.
- Solve problems involving, for example, the instantaneous acceleration of a car moving on a hill with resistance.

Before you start

You should know how to:

1. Use Newton's second law.

e.g. Find the force required to make a car of mass 1000 kg accelerate by 0.5 ms^{-2}, if the car experiences a resistance of 300 N.

Using Newton's second law ($F = ma$), the resultant force is mass times acceleration, which gives

$F - 300 = 1000 \times 0.5$

Hence $F = 800$ N

Skills check:

1. a) A bicycle and rider of mass 75 kg are travelling along a straight horizontal road. The cyclist exerts a forward force of 80 N. The resistance to the motion of the cyclist is 20 N. Find the acceleration of the cyclist.

b) A car of mass 1200 kg is travelling down a road inclined at 10° to the horizontal. The car experiences a total resistance of 800 N and the engine is switched off. Find the acceleration of the car.

2. Find the work done by a force.
   e.g. A child exerts a horizontal force of 80 N on a sledge carrying a load. Find the work done by the child in moving the sledge 55 m.

   Using the definition of work done as force in direction of motion times distance moved ($F \times s$),

   work done = 80 × 55 = 4400 J = 4.4 kJ.

2. a) Two men are pushing a car. They each exert forces of 120 N in the direction of motion of the car. Find the work done by the men in moving the car 25 m.
   b) A constant forward force of 8000 N acts on a lorry as it moves forward on a horizontal road. Find the work done by the force as it moves forward 60 m.
   c) The work done in moving a train forward along a horizontal track is 40 kJ. If the train moves a distance of 80 m, find the constant driving force of the train's engine.

8.1 Power as rate of doing work

The work done by a body is important, as we found out in Chapter 7. In some situations it is not only the work done that is significant, but also the time taken to do that work. For example, when a car manufacturer advertises time taken for their car to reach a given speed, they are stating something about the rate at which the engine can work.

The rate at which work is done is called **power**, which can be expressed in a formula.

$$\text{power} = \frac{\text{work done}}{\text{time}}$$

One unit of power is produced when work is done at the rate of 1 joule per second. This unit is called the watt, W, after **James Watt (1736–1819)**, a Scottish inventor and engineer best known for his work on steam engine development.

For example, if it takes a car 50 seconds to move 100 m when an average force produced by the engine is 300 N, then the average rate of working or average power is

$$\frac{\text{work done}}{\text{time}} = \frac{300 \times 100}{50} = 600\,\text{W}.$$

In some cases when the power is specified, the speed is required at a given time. A car that is moved a distance s metres in t seconds by a driving force F newtons has

$$\text{power} = \frac{F \times s}{t} = F \times \frac{s}{t}.$$

But the quantity $\frac{s}{t}$ is the speed, v ms^{-1}, of the car.

Hence

power = Fv

If the speed is not constant, the value of Fv gives the power at the instant when the speed is v ms^{-1}.

There is obviously a limit to the power that a car or any other vehicle can produce. When the maximum power is attained, the speed produced is also a maximum. In this case there is no acceleration possible and the resultant force is zero.

Example 1

A car moves along a horizontal road against a resistance of 750 N. The maximum power of the car's engine is 12 kW. Find the maximum speed of the car.

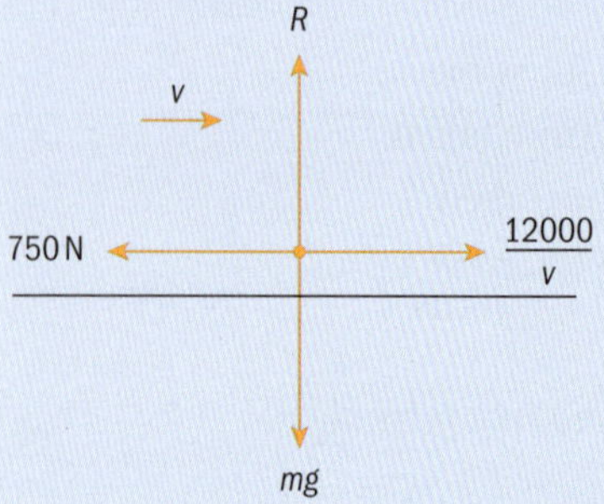

Resolving horizontally $D - 750 = 0$ so $D = 750$

There is no acceleration when at maximum speed so the forces balance.

To achieve maximum speed, maximum power of 12000 W needs to be used.

In this calculation, D is used to represent the driving force.

Maximum power = driving force × maximum speed

$12000 = 750 \times v$

$v = 16$ ms^{-1}

Example 2

A mass of 164 kg is raised, by a crane, vertically upwards through a distance of 35 m in 106 seconds. Find the average power of the crane.

The work done against gravity is $164g \times 35 = 57400$ J

This work is done in 106 s.

Hence the average power is $\frac{57400}{106} = 541.509434 \approx 542$ W (to 3 s.f.)

Exercise 8.1

1. Find the average power in raising a body of mass 45 kg a vertical distance of 80 m in 20 s.

2. Find the average power when lifting a mass of 32 kg vertically at a constant speed of 5 ms^{-1}.

3. Find the average rate at which a climber of mass 90 kg must work when climbing a vertical distance of 24 m in 2 minutes.

4. In building a section of a wall, a man has to lift 400 bricks a vertical distance of 140 cm. Each brick has a mass of 1.8 kg and the man completes the section of wall in 8 minutes. Find the man's average rate of working.

5. A car is driven along a straight horizontal road against a resistance to motion of 600 N. Find the maximum speed of the car when its engine has a power of

   a) 4 kW **b)** 6.8 kW **c)** 9.2 kW.

6. With its engine working at a constant rate of 28 kW, the maximum speed a car can attain on level ground is 36 ms^{-1}. Find the magnitude of the resistances on the car.

7. A cyclist travels along a straight horizontal road at a constant speed of 9 ms^{-1}. The resistance to motion of the cyclist is constant and totals 60 N. Find the power of the cyclist.

8. A motorcyclist is travelling at a constant speed of 28 ms^{-1} along level ground. The power output of the engine is 17500 W. Find the total resistance on the motorcyclist.

9. A cyclist and her bicycle have a total mass of 90 kg. The resistance to her motion is 24 N and the rate at which she is working is 250 W.

   a) Find the maximum speed when she is travelling on a straight horizontal road.

   b) Find the maximum speed when she is travelling up the line of greatest slope of a hill inclined at an angle 2° to the horizontal.

   c) Find the maximum speed when she is travelling down the line of greatest slope of a hill inclined at an angle 0.5° to the horizontal.

10. With its engines working with constant power of 350 kW, a train of mass 250 000 kg climbs a hill inclined at an angle 1° to the horizontal with constant speed of 7 ms^{-1}. Find the magnitude of the resistance to motion experienced by the train.

11. A train of mass 400 tonnes is travelling along a straight horizontal track with a constant speed of $25\,\text{ms}^{-1}$. The train experiences a constant resistance to motion of magnitude 250 000 N.

a) Find the rate at which the train's engine is working.

The train now moves up a hill inclined at an angle θ to the horizontal, where $\sin\theta = \frac{1}{20}$. The engine continues to work at the same rate and the magnitude of the non-gravitational resistance remains the same.

b) Find the new constant speed.

12. A car of mass 2400 kg is travelling at constant speed $18\,\text{ms}^{-1}$ up the line of greatest slope of a road inclined at 8° to the horizontal. The non-gravitational resistance to motion is modelled as a single force of magnitude 600 N.

a) Find the power of the car's engine.

When the car passes a point A, travelling at $18\,\text{ms}^{-1}$, the engine is switched off and the car comes to rest, without braking, a distance s m from A.

Find

b) the distance s

c) the time taken for the car to come to rest.

13. A load of mass 1250 kg is raised by a crane from rest on horizontal ground, to rest at a height 1.54 m above the ground. The work done against the resistance to motion is 5750 J.

i) Find the work done by the crane. [3]

ii) Assuming the power output of the crane is constant and equal to 1.25 kW, find the time taken to raise the load. [2]

Cambridge International AS and A Level Mathematics 9709, Paper 41 Q2 May/June 2011

8.2 Acceleration and variable resistance

If, at a particular instant, a vehicle exerts more driving force than the total resistive forces there will be a resultant force in the direction of motion. In this case, the vehicle will accelerate. The acceleration can be found by applying Newton's second law.

Be aware that the acceleration will be different at different instants. This is because if the power remains constant, the vehicle accelerates, changing its speed and hence the driving force will change. So we can only calculate acceleration at a particular instant in time.

Example 3

A car travels along a horizontal straight road against a constant resistive force of magnitude 275 N. The mass of the car is 1400 kg and its engine is working at a rate of 7.5 kW. Find

a) the acceleration at the instant when the car has a speed of $12\,\text{ms}^{-1}$

b) the speed of the car at the instant when it is accelerating at a rate of $0.1\,\text{ms}^{-2}$ up the line of greatest slope of a hill inclined at an angle of 5° to the horizontal.

a)

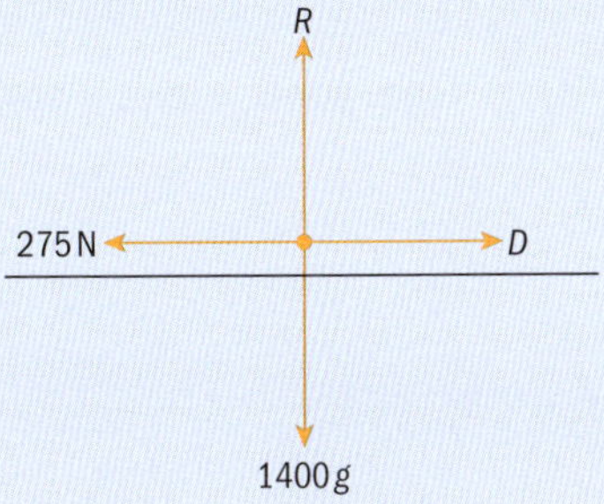

Using $P = Dv$ gives $7500 = D \times 12$, so $D = 625\,\text{N}$.

Resolving horizontally and using $F = ma$ gives $625 - 275 = 1400a$

Hence $a = 0.25\,\text{ms}^{-2}$.

b)

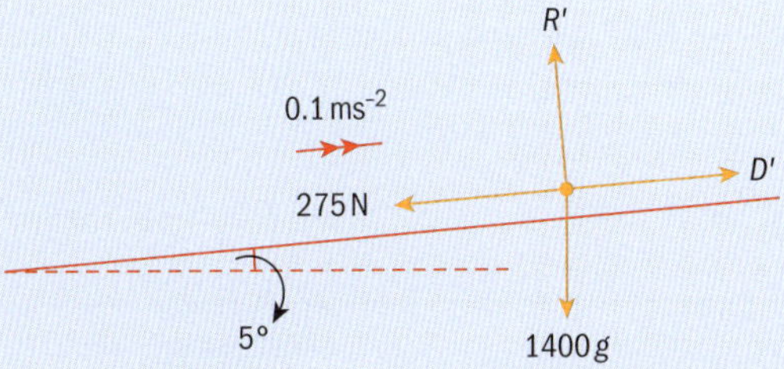

The driving force is $D' = \dfrac{7500}{v}$.

Using $F = ma$ and resolving parallel to the slope gives

$$\frac{7500}{v} - 275 - 1400g\sin 5 = 1400 \times 0.1$$

Hence, $v = 4.586649893 \approx 4.59\,\text{ms}^{-1}$.

Example 4

A car of mass 1250 kg travels on a straight horizontal road. It experiences a resistive force of magnitude 25 v N, where v ms^{-1} is the car's speed. The maximum speed of the car on this road is 60 ms^{-1}. Find

a) the car's maximum power

b) the car's maximum possible acceleration when its speed is 30 ms^{-1}.

a) When the car is travelling at its maximum speed, there is no acceleration and so the driving force of the car, D, must be equal to the resistive force.
Hence $D = 25v$.

Power = $(25v) \times v = 25v^2 = 25 \times 60^2 = 90\,000$ W = 90 kW ← Since power = force × speed

b)

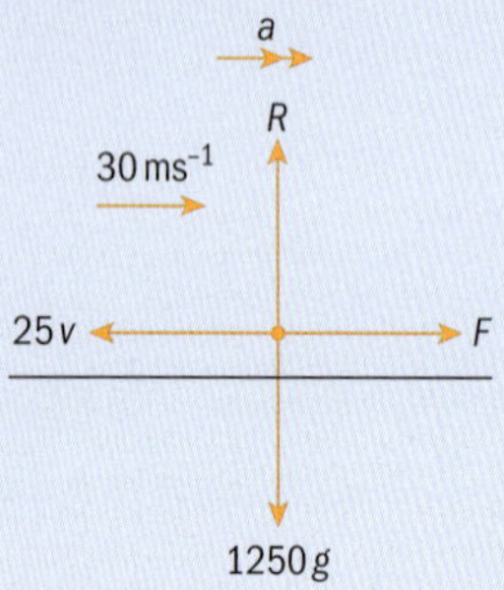

Driving force, $F = \frac{90\,000}{30} = 3000$ N

Resolving forces horizontally and using Newton's second law gives

$3000 - (25 \times v) = ma$

$3000 - (25 \times 30) = 1250\,a$

$a = 1.8$ ms^{-2}

Exercise 8.2

1. A car of mass 1400 kg is travelling along a straight horizontal road. The resistance to motion of the car is 500 N. At the instant that the car's engine is working at a rate of 8 kW the car has a speed of 10 ms^{-1}. Find the car's acceleration at this instant.

2. A car of mass 800 kg is driven along a horizontal road against a constant resistance to motion of 250 N. With the engine of the car working at a rate of 12 kW, find

   a) the acceleration of the car when its speed is 3 ms^{-1}

   b) the acceleration of the car when its speed is 12 ms^{-1}

   c) the maximum speed of the car.

3. A car of mass 1000 kg is driven along a straight horizontal road against a constant resistance to motion of 300 N. With the engine of the car working at a rate of 8 kW, find

   a) the acceleration of the car when its speed is 4 ms^{-1}

   b) the speed of the car when its acceleration is 2 ms^{-2}

   c) the maximum speed of the car.

4. A train of mass 80 tonnes travels along a level track with its engines developing a constant power of 54 kW.

   a) If the greatest speed the train can reach on this track is 30 ms^{-1}, find the magnitude of the resistance to motion.

   b) Assuming the resistance remains constant and the train's engines still work at 54 kW, find the acceleration of the train when it travels along the track at 15 ms^{-1}.

5. a) A car of mass 850 kg has a maximum speed on a straight horizontal road of 40 ms^{-1}. Find the maximum power of the engine if the resistance to motion is 280 N.

   b) The car is travelling at 20 ms^{-1} on a hill, inclined at $\theta°$ to the horizontal, where $\sin\theta = \frac{1}{40}$. The resistance to motion is constant and the engine is exerting maximum power. Find the acceleration of the car.

6. A lorry of mass 8000 kg has a maximum speed of 24 ms^{-1} on a straight horizontal road. The maximum power of the lorry's engine is 30 kW.

   a) Find the total resistance to motion at this speed.

   It is given that the resistance to motion is kv^2, where v ms^{-1} is the speed of the lorry and k is a constant.

   b) Find the value of k.

   The lorry now climbs a hill inclined at 2° to the horizontal.

   c) When the lorry is travelling at 12 ms^{-1} and accelerating at a rate of 0.25 ms^{-2}, find the power exerted by the engine.

7. A car of mass 1200 kg moves up a road inclined at an angle θ to the horizontal, where $\sin\theta = \frac{1}{20}$. The car's engine produces a constant power of 30 kW. The car experiences a resistive force of magnitude kv N, where k is a constant and v is the car's speed in ms^{-1}. At the instant the car has speed $10\,ms^{-1}$ its acceleration is $0.8\,ms^{-2}$.

   a) Show that $k = 144$

   b) Show that V, the maximum speed of the car up this road, satisfies

   $6V^2 + 25V - 1250 = 0.$

   Hence find the maximum speed of the car on this road.

8. A cyclist and her bicycle have a total mass of 50 kg. She produces a maximum power of 80 W. The resistance to the motion of the cyclist is proportional to her speed. On a straight horizontal road she can travel at a maximum of $10\,ms^{-1}$. On a slope inclined at $\alpha°$ to the horizontal, she can freewheel down the slope with a maximum speed of $15\,ms^{-1}$. Find

   a) the value of α

   b) the maximum speed at which she can go up the same slope.

9. A car's engine is working with maximum power P W. The car has a mass of 1000 kg and is travelling along a straight level road and experiences a resistance of R N. At one instant it is travelling with speed $4\,ms^{-1}$ with an acceleration of $2.5\,ms^{-2}$. At a later time, it is travelling with speed $8\,ms^{-1}$ with an acceleration of $0.5\,ms^{-2}$. Find the values of P and R.

10. A car's engine has a power output of 50 kW and the car has a mass of 1000 kg. On a straight horizontal road, the car has a greatest speed of $40\,ms^{-1}$. The resistance to the cars speed is variable and has magnitude kv N, where k is a constant and v is the car's speed in ms^{-1}.

   a) Find k.

   The car now travels on a road inclined at 5° to the horizontal.

   b) Find the maximum speed of the car when going up the hill.

   c) Find the maximum speed of the car when going down the hill.

11. A car travels along a horizontal straight road with increasing speed until it reaches its maximum speed of 30 ms^{-1}. The resistance to motion is constant and equal to R N, and the power provided by the car's engine is 18 k W.

i) Find the value of R. [3]

ii) Given that the car has mass 1200 kg, find its acceleration at the instant when its speed is 20 ms^{-1}. [3]

Cambridge International AS and A Level Mathematics 9709, Paper 4 Q3 May/June 2007

12. A car of mass 600 kg travels along a horizontal straight road, with its engine working at a rate of 40 kW. The resistance to motion of the car is constant and equal to 800 N. The car passes through the point A on the road with speed 25 ms^{-1}. The car's acceleration at the point B on the road is half its acceleration at A. Find the speed of the car at B. [5]

Cambridge International AS and A Level Mathematics 9709, Paper 41 Q2 November 2010

Summary exercise 8

1. A car of mass 1150 kg travels up a straight hill inclined at 1.2° to the horizontal. The resistance to motion of the car is 975 N. Find the acceleration of the car at an instant when it is moving with speed 16 ms^{-1} and the engine is working at a power of 35 kW. [4]

Cambridge International AS and A Level Mathematics 9709, Paper 41 Q1 May/June 2010

2. A car has a maximum speed of 50 ms^{-1} on a straight horizontal road and a maximum power output of 40 kW. The resistive force, R N at a speed of v ms^{-1}, is given by $R = kv$. Find

a) the value of k

b) the resistance when the car's speed is 20 ms^{-1}

c) the power needed to travel at a constant speed of 20 ms^{-1} along a straight horizontal road.

EXAM-STYLE QUESTION

3. a) The power produced by the engine of a car, as it travels on a straight horizontal road at a constant speed of 60 ms^{-1}, is 30 kW. Find the resistance to motion of the car.

b) The car has a mass of 1200 kg and is ascending a hill inclined at 4° to the horizontal. Given that the rate at which the car's engine is working is unchanged and the resistance to motion is constant, find the maximum steady speed on the hill.

4. A train of mass 400 000 kg is moving on a straight horizontal track. The power of the engine is constant and equal to 1500 kW and the resistance to the train's motion is 30 000 N. Find

   i) the acceleration of the train when it speed is $37.5\,\text{ms}^{-1}$, [4]

   ii) the steady speed at which the train can move. [2]

Cambridge International AS and A Level Mathematics 9709, Paper 41 Q4 May/June 2013

5. A car of mass 900 kg travels along a horizontal straight road with its engine working at a constant rate of P kW. The resistance to motion of the car is 550 N. Given that the acceleration of the car is $0.2\,\text{ms}^{-2}$ at an instant when its speed is $30\,\text{ms}^{-1}$, find the value of P. [4]

Cambridge International AS and A Level Mathematics 9709, Paper 4 Q1 October/November 2007

EXAM-STYLE QUESTION

6. A car has a mass of 800 kg and a maximum power of P W. The car has a top speed of $45\,\text{ms}^{-1}$ on a straight horizontal road and a top speed of $24\,\text{ms}^{-1}$ up a hill inclined at an angle θ to the horizontal, where $\sin\theta = 0.4$. Given that the resistance to motion is constant and has magnitude R N, write down two equations connecting P and R. Hence find the values of P and R.

7. The forces resisting the motion of a car are constant at all speeds and total 480 N. When the engine is working at a rate of P kW, the maximum speed of the car on a straight horizontal road is $36\,\text{ms}^{-1}$.

   a) Find the value of P.

   The car is moving at this maximum speed when the power of engine is suddenly increased to $(P + 20)$ kW and the resulting initial acceleration of the car is $0.25\,\text{ms}^{-2}$.

   b) Find the mass of the car.

   The car travels down a road a straight road inclined at 6° to the horizontal at a constant speed of $12\,\text{ms}^{-1}$ with the engine working at a rate of Q kW to provide a constant braking force.

   c) Find the value of Q.

EXAM-STYLE QUESTION

8. The total mass of a motorbike and its rider is 300 kg. The maximum power of the motorbike's engine is P kW. When the speed of the motorbike is $v\,\text{ms}^{-1}$, the resistance to motion is kv N, where k is a constant. The motorbike has a maximum steady speed of $20\,\text{ms}^{-1}$ when ascending a hill inclined at an angle $\sin^{-1}\frac{1}{10}$ to the horizontal.

   a) Show that $5P = 2k + 30$.

   The maximum steady speed going down the same hill is $30\,\text{ms}^{-1}$.

   b) Find another equation relating k and P and hence find the values of k and P.

   c) Find the maximum steady speed of the motorbike on a straight horizontal road.

9. A car of mass 1200 kg moves in a straight line along horizontal ground. The resistance to motion of the car is constant and has magnitude 960 N. The car's engine works at a rate of 17 280 W.

i) Calculate the acceleration of the car at the instant when its speed is $12\,\text{ms}^{-1}$. [3]

The car passes through points *A* and *B*. While the car is moving between *A* and *B* it has constant speed $V\,\text{ms}^{-1}$.

ii) Show that $V = 18$. [2]

At the instant when the car reaches *B* the engine is switched off and subsequently provides no energy. The car continues along the straight line until it comes to rest at the point *C*. The time taken for the car to travel from *A* to *C* is 52.5s.

iii) Find the distance AC. [5]

Cambridge International AS and A Level Mathematics 9709 Paper 41 Q7 October/November 2012

10. A car of mass 1200 kg is travelling on a horizontal straight road and passes through a point A with speed $25\,\text{ms}^{-1}$.

The power of the car's engine is 18 kW and the resistance to the car's motion is 900 N.

i) Find the deceleration of the car at A. [4]

ii) Show that the speed of the car does not fall below $20\,\text{ms}^{-1}$ while the car continues to move with the engine exerting a constant power of 18 kW. [2]

Cambridge International AS and A Level Mathematics 9709, Paper 4 Q3 October/November 2008

Chapter summary

- Power is the rate at which work is done and is given by $P = \dfrac{\text{work done}}{\text{time}}$.
- The power produced by a force F on a vehicle moving with speed $v\,\text{ms}^{-1}$ is given by $P = Fv$.
- The S.I. unit of power is the watt (W).

9 Variable forces

In the 19th century, Lord Kelvin proposed an analogue machine that would mechanically perform the processes of integration. Vannevar Bush built the first successful machine in 1930 in America, and in 1935, Douglas Hartree built the differential analyzer at Manchester University. Hartree's machine was able to perform the calculations required for a number of purposes. During the Second World War, for example, British scientists were able to calculate the trajectories of V2 rockets with it. These analogue machines were the forerunners of modern digital computers that can tackle these calculations with ease.

Objectives

- Use differentiation and integration with respect to time to solve simple problems concerning displacement, velocity and acceleration.

Before you start

You should know how to:

1. Differentiate polynomial functions in x^n (for any rational n), e.g.

$$\frac{d}{dx}\left(3x^2 - \frac{4}{x} + 6x^{\frac{1}{2}}\right) = 6x + \frac{4}{x^2} + 3x^{-\frac{1}{2}}$$

2. Locate stationary points.

e.g. find the stationary points of the curve $y = 2x^3 + 3x^2 - 36x + 4$

$\frac{dy}{dx} = 6x^2 + 6x - 36$ so at the stationary points

$$6x^2 + 6x - 36 = 0$$
$$x^2 + x - 6 = 0$$
$$(x + 3)(x - 2) = 0$$
$$x = -3 \text{ or } x = 2$$

Hence the stationary points are $(-3, 85)$ and $(2, -40)$

3. Integrate $(ax + b)^n$ (for any rational n except -1)

e.g. $\int (2x + 1)^{\frac{1}{2}}\, dx = \frac{2}{3}(2x + 1)^{\frac{3}{2}}\left(\frac{1}{2}\right) + c$

$= \frac{1}{3}(2x + 1)^{\frac{3}{2}} + c$

Skills check:

1. Differentiate with respect to x

a) $x^2 - 2x$ b) $\frac{1}{x} - \frac{1}{x^2}$ c) $\sqrt{x}$

2. Find the maximum and minimum points of the function $y = 9x^2 - 4x^3$

3. Integrate with respect to x

a) $3x^2 + 4x$ b) $\frac{1}{2x^2}$ c) $\sqrt{x}$

d) $(1 + 2x)^3$ e) $\sqrt{2x - 1}$

4. Solve problems involving evaluating a constant of integration.

e.g. if $\frac{dy}{dx} = 3x^2 - 4x$ and $y = 3$ when $x = 1$,

$y = \int 3x^2 - 4x \, dx$

$y = x^3 - 2x^2 + c$

$3 = 1 - 2 + c$

$c = 4$

$y = x^3 - 2x^2 + 4$

4. Integrate $y = 2x^2 - x - 1$ with respect to x. If the value of the integral is -1 when $x = 1$, find the value of the constant of integration.

5. Evaluate definite integrals.

e.g. $\int_4^9 3x^{\frac{1}{2}} dx = \left[2x^{\frac{3}{2}}\right]_4^9 = (2 \times 27) - (2 \times 8) = 38$

5. Find $\int_1^2 6x^2 + 2x \, dx$

9.1 Using differentiation to describe straight line motion

Earlier in this book, we looked at straight line motion We studied situations where either velocity was constant or, as force is directly proportional to acceleration, situations where the force applied to an object was constant too, such as with gravitational force.

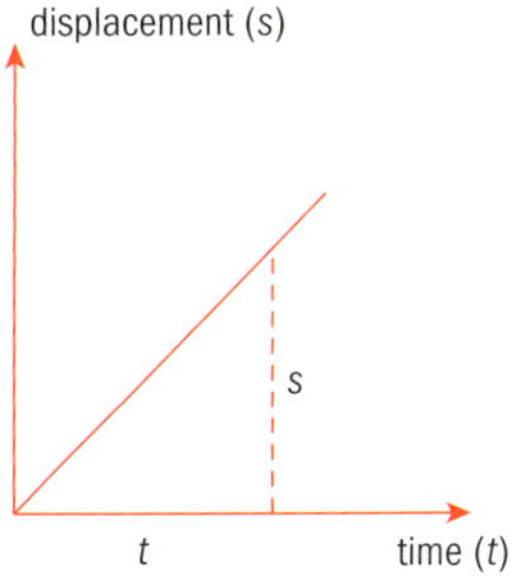

We know that velocity $= \frac{\text{displacement}}{\text{time}}$ so velocity $= \frac{s}{t}$

velocity = gradient of displacement-time graph

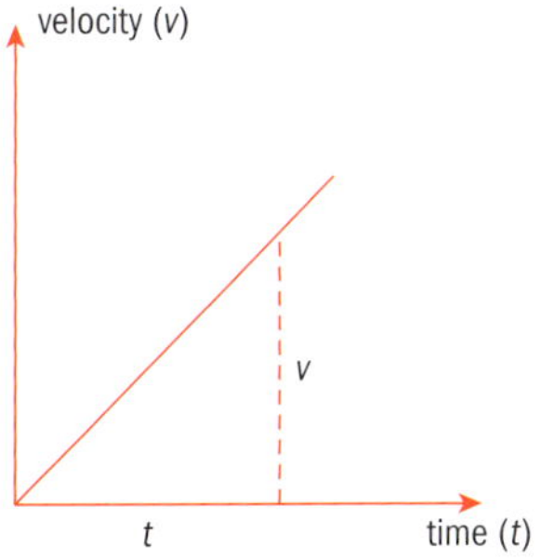

We know that acceleration $= \frac{\text{velocity}}{\text{time}}$ so acceleration $= \frac{v}{t}$

acceleration = gradient of ~~displacement~~ velocity-time graph

Often in real-life situations, acceleration is neither constant, nor does it change suddenly. If you drove out of your garage in the morning and kept accelerating, then you would eventually find yourself travelling beyond the speed limit. If the elevator in your tower block changed suddenly from accelerating upwards to decelerating you would fall over as a result of the jerk and the lift cable would be likely to break. In reality, the acceleration of objects is changing with time.

If we looked at a velocity time graph, the line would be curved and not straight.

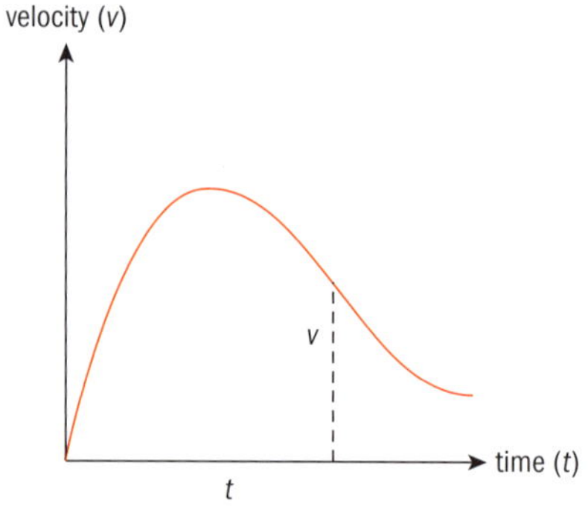

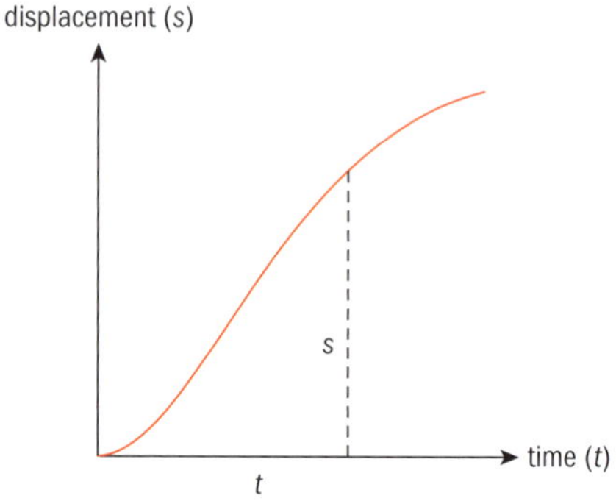

Acceleration is defined as the rate of change of velocity with respect to time. This is the gradient of the curve above. In mathematical terms, we measure the rate at which velocity v changes with respect to time t as the **derivative** $\frac{dv}{dt}$. For constant acceleration this is equivalent to our definition of acceleration as the gradient of the velocity–time graph. Similarly, velocity is defined as the rate of change of displacement with respect to time. So the rate at which displacement s changes with respect to time t is $\frac{ds}{dt}$.

Note: It is very important to remember that the constant acceleration formulae that you learned in Chapter 2 should only be used where a is a *constant*. Trying to apply these formulae in problems involving variable acceleration will lead to error.

For a particle moving in a straight line: $a = \frac{dv}{dt}$ and $v = \frac{ds}{dt}$.

Example 1

A car accelerates to overtake a truck on the highway. After passing the truck, the car slows down as it approaches slow-moving traffic, finally coming to a halt. The velocity $v\,\text{ms}^{-1}$ of the car after time t seconds is given by the equation $v = 8 + 0.36t^2 - 0.02t^3$. Find how fast the car is moving when it reaches its maximum velocity.

The acceleration of the car can be found by differentiating v:

$$a = \frac{dv}{dt} = 0.72t - 0.06t^2$$

When the car reaches its maximum velocity, the acceleration is momentarily equal to zero.

$$0.72t - 0.06t^2 = 0$$
$$0.06t\,(12 - t) = 0$$
$$t = 0 \text{ or } t = 12$$

The car reaches its maximum velocity after 12 s. To find the velocity, substitute $t = 12$ back into the equation for velocity.

$$v = 8 + 0.36 \times 144 - 0.02 \times 1728$$
$$v = 25.28$$

The maximum velocity is $25.3\,\text{ms}^{-1}$ (to 3 s.f.)

Example 2

A particle moves on a straight line through O so that its displacement s from O at time t seconds is given by the equation $s = 4 + 12t^2 - t^3$. Find the total distance covered in the first 12 seconds. What will be the acceleration of the particle when it is momentarily at rest?

Distance and displacement are different when in different directions, so we need to check when the particle changes direction. This will occur when $v = 0$.

$$v = \frac{ds}{dt} = 24t - 3t^2 = 3t(8 - t)$$

$v = 0$ when $t = 0$ or when $t = 8$

When $t = 6$, $v = 36$ and when $t = 10$, $v = -60$. Since the velocities have opposite signs we can conclude that the particle changes direction in between, when $t = 8$.

When $t = 0$, $s = 4$, when $t = 8$, $s = 260$ and when $t = 12$, $s = 4$.

From this we can see that the total *displacement* between $t = 0$ and $t = 12$ is 0 as the particle returns to the point where it started. The total distance covered is

$$2 \times (260 - 4) = 512\,\text{m}$$

$$a = \frac{dv}{dt} = 24 - 6t \text{ so when } t = 8,\ a = 24 - 6 \times 8 = -24\text{ms}^{-2}$$

Exercise 9.1

1. The velocity of a particle at time t is given by $v = 2t^2 - 4t + 3$. Find the time at which the acceleration is zero.

2. The velocity of a particle at time t is given by $v = 5 + 4t - 3t^2$. Find the velocity when the acceleration is zero.

3. The velocity, v of a particle at time t is given by $v = t^3 - 6t^2 + 9t$. Find an expression for a and the maximum velocity of the particle.

4. The displacement, s of a particle from a point O at time t is given by the equation $s = t(2t - 1)(t + 1)$. Find expressions for the velocity and the acceleration of the particle.

5. The displacement s of a particle from a point O at time t is given by the equation $s = 2t^4 - 27t$. Find expressions for the velocity and acceleration of the particle. Find the acceleration at the instant that the velocity is zero.

6. The displacement s of a particle from a point O at time t is given by the equation $s = 2t^3 - 3t^2 - 72$. Find expressions for the velocity and the acceleration of the particle. Find the displacement of the particle at the points where it is momentarily at rest.

7. The velocity, v of a particle at time t is given by $v = t^2 + \frac{1}{4t}$. Find an expression for the acceleration of the particle. Find the time when the acceleration is zero and the velocity at that instant.

8. The displacement s m of a particle from a point O at time t seconds is given by the equation. $s = 3\sqrt{t} - \frac{1}{2}t$. Find

   a) the time of maximum displacement from O

   b) the time when the velocity of the particle is $0.25\,\text{ms}^{-1}$ and the acceleration at that instant.

9. Between two stations, the distance s of a train from the first station is given by the equation $s = 8t^3 - 6t^4$ km/h. The train first comes to rest at its destination. Find

   a) the time that the journey between the two stations takes

   b) the distance between the two stations

   c) the maximum speed that the train reaches.

10. A particle moves along a straight line though a point O from $t = 0$ until $t = 3$. The displacement s of the particle at time t is given by the equation $s = 2t^3 - 9t^2 + 12t - 4$. Find:

    a) an expression for the velocity of the particle

    b) the times when the particle changes direction

    c) the position of the particle at these points

    d) the total distance travelled by the particle between $t = 0$ and $t = 3$.

9.2 Using integration to describe straight line motion

In **9.1** we discovered how to find velocity when we know displacement and how to find acceleration when we know velocity. In Chapter 1 we saw that when we wish to find displacement from knowing velocity or to find velocity from knowing acceleration then this was done by finding the area under a graph for constant velocity or acceleration.

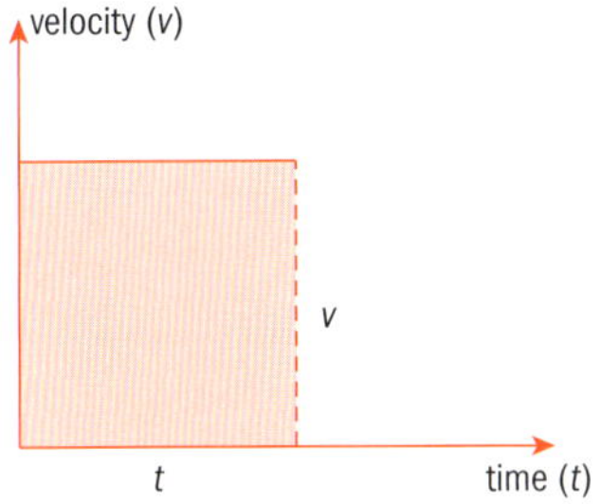

When velocity is constant,

displacement = velocity × time hence displacement = $v \cdot t$

displacement is found by calculating the area of the rectangle.

When acceleration is constant, the velocity graph will be a straight line. Consider the area under the graph to be made of a series of very narrow rectangles. The area of each of these rectangles is the displacement in a very short time. Combining these areas, we get an approximation to the area of the trapezium under the graph, which improves as the time period for these rectangles decreases.

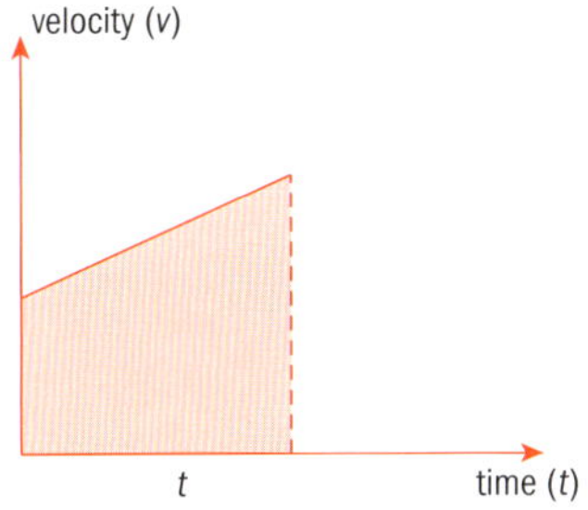

displacement = area under displacement–time graph

In a similar way we found that velocity was the area of a rectangle under an acceleration–time graph when acceleration was constant.

We now look the situation when acceleration is not constant. As before we will have a curved graph, either a curved velocity–time graph or a curved acceleration–time graph.

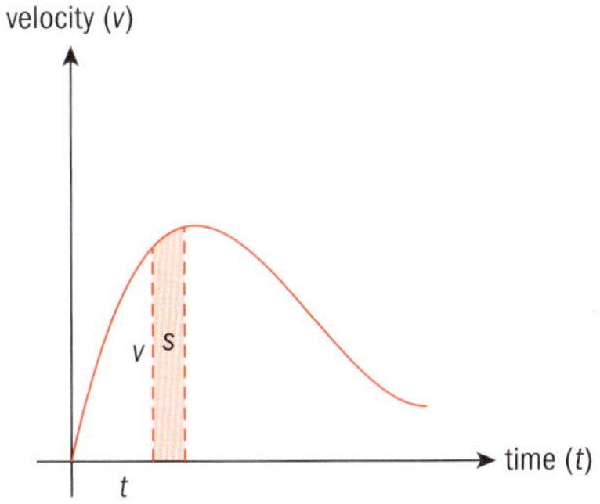

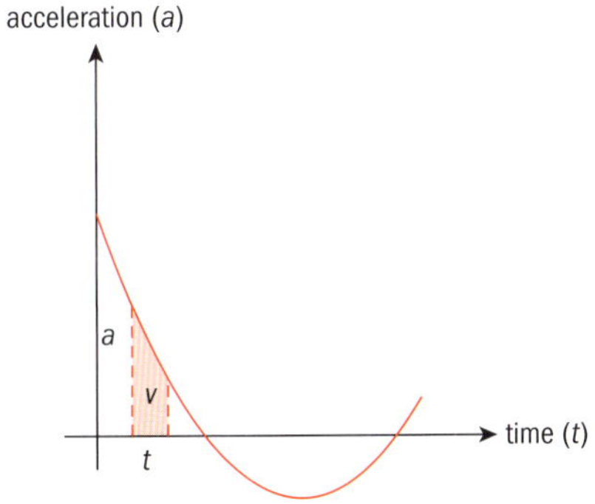

Finding the area under a curve is known in calculus as anti-differentiation or integration. We will use integration to describe the processes of finding displacement when we know velocity and finding velocity when we know acceleration.

In the following examples you will see different situations where you will need to use indefinite integrals (integrals that have a constant of integration in the answer) and definite integrals (integrals that are found between limits).

Example 3

If the velocity of a particle that moves along a straight line through a point O at time t is given by the equation $v = 2t^3 - 3t^2$ and the displacement of the particle from O when $t = 2$ is 10 m, then find an expression for displacement.

In this example we are given a particular value of the displacement function for a given value of time. We use an indefinite integral.

Integrating

$$s = \int 2t^3 - 3t^2 \, dt$$

$$s = \frac{1}{2}t^4 - t^3 + c$$

when $t = 2$, $s = 8 - 8 + c = 10$

$c = 10$

$$s = \frac{1}{2}t^4 - t^3 + 10$$

Example 4

If the velocity of a particle that moves along a straight line through O is given by the equation $v = 5t - \frac{1}{2}t^2$, find the distance travelled between $t = 2$ and $t = 4$.

In this example we do not know the displacement from a fixed point for any value of t. Since we are finding the value between two values of t we use a definite integral.

Integrating between the limits $t = 2$ and $t = 4$

$$s = \int_2^4 5t - \frac{1}{2}t^2 \, dt$$

$$s = \left[\frac{5}{2}t^2 - \frac{1}{6}t^3\right]_2^4$$

$$s = \left(40 - \frac{32}{3}\right) - \left(10 - \frac{4}{3}\right)$$

$s = 20.7$ m

For a particle moving in a straight line, displacement, velocity and acceleration can be found from each other as summarized in this diagram:

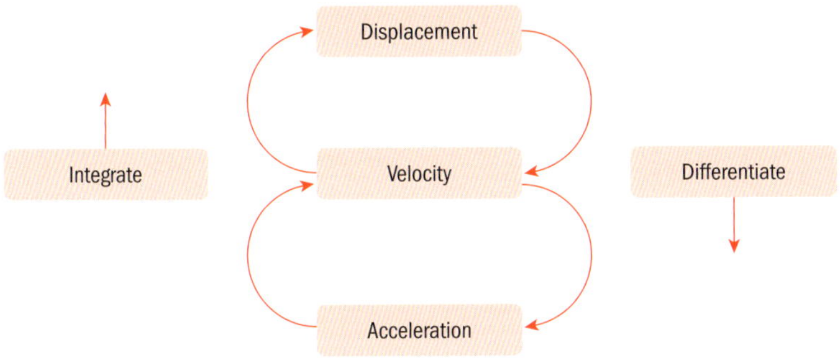

Exercise 9.2

1. The velocity v of a particle at time t is given by $v = 8t^3 - 6t^2 + 5$.
If the particle starts at a point O, find an expression, in terms of t for its displacement from O.

2. If the velocity of a particle at time t is given by $v = 15t^4 - 12t^2$ and the displacement when $t = 2$ is 34 m, then find an expression for displacement and find the displacement of the particle when $t = 1$.

3. The acceleration of a particle that moves along a straight line through a point O is given by $a = 12 - 3t^2$. When $t = 2$, the velocity of the particle is $8\,\text{ms}^{-1}$ and its displacement from O is 2 m. Find expressions for the velocity and displacement of the particle.

4. If the velocity of a particle is given by $v = 20t - 6t^2$, find the distance travelled between $t = 1$ and $t = 3$.

5. A particle moves from rest in a straight line from a point O with acceleration $a = 3t^2 + 4t - 5$. Find its displacement between $t = 1$ and $t = 2$.

6. A car moves away from traffic lights with acceleration $a = 2t - t^2$ and initial velocity $v = 3$. Find an expression for the velocity of the car and its maximum velocity. Find the distance travelled before the car reaches its maximum velocity.

7. A particle moves along a straight line through a point O so that its velocity at time t is given by $v = 4\,(1 + 2t)^3$. Find an expression for s, given that $s = 2$ when $t = -1$.

8. The acceleration of a particle that moves in a straight line is given by the equation $a = 15\sqrt{2t - 1}$. Find expressions for velocity and displacement, given that when $t = \frac{1}{2}$, $v = 0$ and $s = 0$.

9. The velocity of a particle that moves along a straight line through O is given by the formula $v = 2t - \frac{2}{t^2}$, $t \geq 1$. Given that $s = 3$ when $t = 1$, find expressions for the displacement of the particle from O and its acceleration.

10. The velocity of a particle that moves along a straight line through O is given by the formula $v = 3t - 2t^2$. When $t = 0$, $s = 0$. Find the maximum velocity and the maximum displacement of the particle from O.

9.3 The constant acceleration formulae

At this point it is interesting to see that the constant acceleration formulae introduced in Chapter 2 are a special case of motion in a straight line, and the equations can be found by using the calculus techniques we have just been learning.

First we should define the parameters and variables we will use.

- a is a constant, equal to the constant acceleration of the particle.
- u is a constant equal to the initial velocity of the particle.
- s is the displacement of the particle from a fixed point O at time t.
- v is the velocity of the particle at time t.

Consider the motion of a particle that starts at point O with an initial velocity u.

$$\frac{dv}{dt} = a$$

$$\int dv = \int a \, dt$$

$$v = at + c_1$$

But when $t = 0$, $v = u$ so $c_1 = u$.

This gives the velocity–time formula $v = u + at$.

$$\frac{ds}{dt} = u + at$$

$$\int dx = \int u + at \, dt$$

$$s = ut + \frac{1}{2}at^2 + c_2$$

But when $t = 0$, $x = 0$ so $c_2 = 0$.

This gives the displacement–time formula $\boxed{s = ut + \frac{1}{2}at^2}$

Other equations can be derived from these by eliminating a or t.

Note: s is used here to denote displacement, but x can also be used.

Summary exercise 9

1. A particle starts from rest from a point O and moves in a straight line. Its speed v ms^{-1} at time t seconds after leaving O is defined as follows.

   For $0 \le t \le 10$, $v = 0.4t - 0.002t^3$

   For $t \ge 10$, $v = \dfrac{2000}{t^2}$

   a) Find the maximum speed of the particle between $t = 0$ and $t = 10$.

   b) Find the total distance travelled in the first 20 s of motion.

2. A car travels from A to B starting from rest. The car's speed increases to a maximum and then it slows down until it is at rest at B. Its velocity (in ms^{-1}) t seconds after leaving A is $0.0000012(240t^4 - 8t^5)$.

   a) Find the distance AB.

   b) Find the time when the acceleration of the car is zero.

   c) Find the maximum speed of the car.

3. A particle starts from a point O and moves in a straight line until it comes to rest so that its velocity t seconds after leaving O is given by $v = \dfrac{6}{(0.2t + 1)^2} - 1$ ms^{-1}.

   a) Find the initial velocity of the particle.

   b) Find the time when it comes to rest.

   c) Find the total distance travelled by the particle.

4. A particle P starts from a point O and moves along a straight line. P's velocity t s after leaving O is vm s^{-1}, where

   $v = 0.16t^{\frac{3}{2}} - 0.016t^2$.

   P comes to rest instantaneously at the point A.

   i) Verify that the value of t when P is at A is 100. [1]

   ii) Find the maximum speed of P in the interval $0 < t < 100$. [4]

   iii) Find the distance OA. [3]

   iv) Find the value of t when P passes through O on returning from A. [2]

   Cambridge International AS and A Level Mathematics 9709, Paper 41 Q7 October/November 2011

5. Two particles P and Q are travelling along a straight line through a point O. P is decelerating at 1 ms^{-2} and at time $t = 0$, its velocity is 3 ms^{-1} and its displacement from O is 2 m. Q starts from the same point with an initial velocity of 1 ms^{-1} and it is decelerating at a rate of 0.2 ms^{-2}. Find when the two particles meet again, how far from O they are and their velocities.

6. A particle P starts at the point O and travels in a straight line. At time t seconds after leaving O the velocity of P is vm s^{-1}, where $v = 0.75t^2 - 0.0625t^3$. Find

   i) the positive value of t for which the acceleration is zero, [3]

   ii) the distance travelled by P before it changes its direction of motion. [5]

   Cambridge International AS and A Level Mathematics 9709, Q4 Paper 41 May/June 2012

Chapter summary

- For a particle moving in a straight line, displacement, velocity and acceleration can be found from each other as summarised in this diagram:

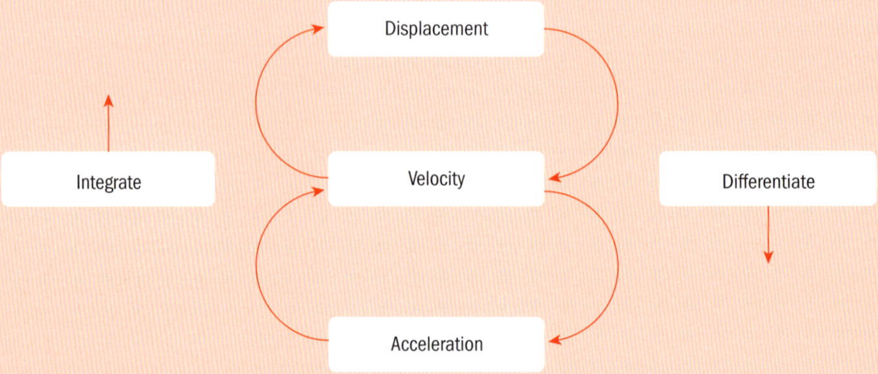

- For a particle moving in a straight line: $a = \frac{dv}{dt}$ and $v = \frac{ds}{dt}$.
- For a particle moving in a straight line: $v = \int a \, dt$ and $s = \int v \, dt$
- Between time t_1 and t_2 The change in velocity is found from the *definite* integral $v = \int_{t_1}^{t_2} a \, dt$ and the displacement is found from the *definite* integral $s = \int_{t_1}^{t_2} v \, dt$

Review exercise C

1. A boy and sledge of total mass 30 kg, are pushed from rest with a horizontal force of 50 N against a resistance of 20 N for 16 m along horizontal ground.

   a) Find his speed after moving this distance.

   The pushing force is now removed and he slows down under the same resistance of 20 N.

   b) Find the total distance travelled from the instant the pushing began until he comes to rest.

2. A man pulls a block, of mass 25 kg, across a horizontal floor by means of a rope attached to the block and inclined at 20° to the horizontal. The block is pulled a distance of 12 m and the tension in the rope is 80 N.

   a) Find the work done by the man.

   The resistance to the motion of the block is 28 N and the block is initially at rest.

   b) Find the final speed of the block.

3. A boy of mass 40 kg is sitting at the top of a water slide at a height of 2.5 m above a swimming pool. The slide is smooth. The boy starts to move down the slide. Find his speed at the instant he enters the pool.

4. A girl of mass 35 kg is at the top of a slide of length 4 m and at a height of 2 m above the ground. She starts to move down the slide from rest. The resistance to motion is constant and of magnitude 20 N. Find her speed at the bottom of the slide.

5. A car of mass 1000 kg descends down the line of greatest slope of a hill inclined at an angle α to the horizontal where $\sin\alpha = 0.2$. The constant resistance to motion has magnitude 300 N. Find the constant breaking force needed to bring the car to rest from $20\,\text{ms}^{-1}$ in a distance of 100 m.

6. 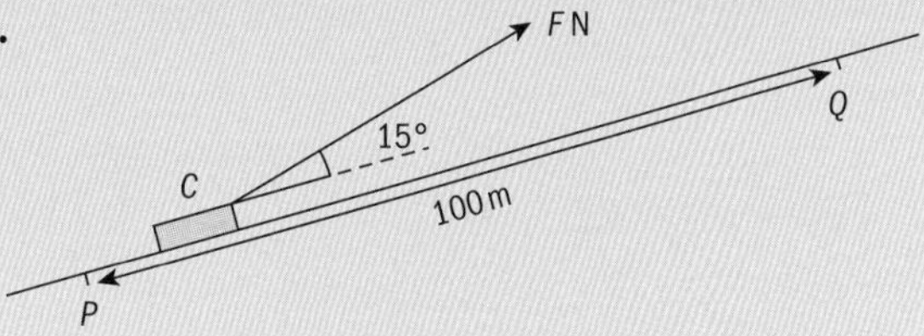

   A crate C is pulled at constant speed up a straight inclined path by a constant force of magnitude F N, acting upwards at an angle of 15° to the path. C passes through points P and Q which are 100 m apart (see diagram). As C travels from P to Q the work done against the resistance to C's motion is 900 J, and the gain in C's potential energy is 2100 J. Write down the work done by the pulling force as C travels from P to Q, and hence find the value of F.

   Cambridge AS and A Level Mathematics 9709, Paper 4 Q2 May/June 2009

7. A particle of mass 0.8 kg slides down a rough inclined plane along a line of greatest slope AB. The distance AB is 8 m. The particle starts at A with speed $3\,\text{ms}^{-1}$ and moves with constant acceleration $2.5\,\text{ms}^{-2}$.

   i) Find the speed of the particle at the instant it reaches B.

   ii) Given that the work done against the frictional force as the particle moves from A to B is 7 J, find the angle of inclination of the plane.

   When the particle is at the point X its speed is the same as the average speed for the motion from A to B.

   iii) Find the work done by the frictional force for the particle's motion from A to X.

   Cambridge AS and A Level Mathematics 9709, Paper 41 Q5 October/November 2010

8. 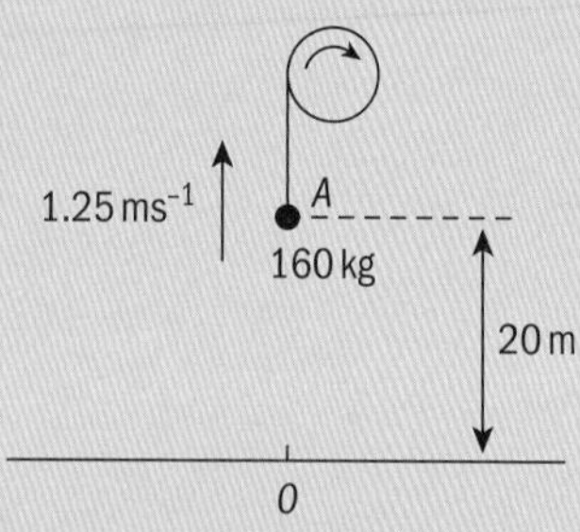

A load of mass 160 kg is pulled vertically upwards, from rest at a fixed point *O* on the ground, using a winding drum. The load passes through a point *A*, 20 m above *O*, with a speed of 1.25 ms^{-1} (see diagram).

Find, for the motion from *O* to *A*,

i) the gain in the potential energy of the load,

ii) the gain in the kinetic energy of the load.

The power output of the winding drum is constant while the load is in motion.

iii) Given that the work done against the resistance to motion from *O* to *A* is 20 kJ and that the time taken for the load to travel from *O* to *A* is 41.7 s, find the power output of the winding drum.

Cambridge AS and A Level Mathematics 9709, Paper 41 Q3 May/June 2012

9. A lorry of mass 15 000 kg climbs from the bottom to the top of a straight hill, of length 1440 m, at a constant speed of 15 ms^{-1}. The top of the hill is 16 m above the level of the bottom of the hill. The resistance to motion is constant and equal to 1800 N.

i) Find the work done by the driving force.

On reaching the top of the hill the lorry continues on a straight horizontal road and passes through a point *P* with speed 24 ms^{-1}. The resistance to motion is constant and is now equal to 1600 N. The work done by the lorry's engine from the top of the hill to the point *P* is 503 kJ.

ii) Find the distance from the top of the hill to the point *P*.

Cambridge AS and A Level Mathematics 9709, Paper 41 Q5 November 2013

10. Find the average rate at which a climber, of mass 75 kg, must work when climbing a vertical distance of 40 m in 150 seconds.

11. A crane lifts a block, of mass 50 kg, to a height of 12 m at a constant speed of 0.6 ms^{-1}. Find the power required.

12. A train, of mass 20 tonnes, produces a maximum power of 2000 kW on a horizontal track. The resistance to the motion of the train is 40 kN.

a) Find the maximum speed of the train.

b) Find the acceleration of the train at the instant it is moving with speed 25 ms^{-1}. Assume the train is working at maximum power.

13. A car of mass 800 kg has a maximum speed of 28 ms^{-1} when travelling up a line of greatest slope of a hill against a resistance of 500 N. The hill is inclined at an angle θ to the horizontal, where $\sin\theta = \frac{1}{40}$. Find the power output of the car's engine.

14. A car of mass 1.4 tonnes moves with constant speed $6\,\text{ms}^{-1}$ up a line of greatest slope of a hill inclined at an angle θ to the horizontal, where $\sin\theta = \frac{1}{7}$. Given that the engine is working at a rate of 18 kW, find the resistance to the motion of the car.

15. A car of mass 880 kg travels along a straight horizontal road with its engine working at a constant rate of P W. The resistance to motion is 700 N. At an instant when the car's speed is $16\,\text{ms}^{-1}$ its acceleration is $0.625\,\text{ms}^{-2}$. Find the value of P.

Cambridge AS and A Level Mathematics 9709, Paper 41 Q1 May/June 2012

16. A load of mass 1250 kg is raised by a crane from rest on horizontal ground, to rest at a height of 1.54 m above the ground. The work done against the resistance to motion is 5750 J.

i) Find the work done by the crane.

ii) Assuming the power output of the crane is constant and equal to 1.25 kW, find the time taken to raise the load.

Cambridge AS and A Level Mathematics 9709, Paper 41 Q2 May/June 2011

17. A block is being pulled along a horizontal floor by a rope inclined at 20° to the horizontal. The tension in the rope is 851 N and the block moves at a constant speed of $2.5\,\text{ms}^{-1}$.

i) Show that the work done on the block in 12 s is approximately 24 kJ

ii) Hence find the power being applied to the block, giving your answer to the nearest kW.

Cambridge AS and A Level Mathematics 9709, Paper 4 Q2 May/June 2008

18.

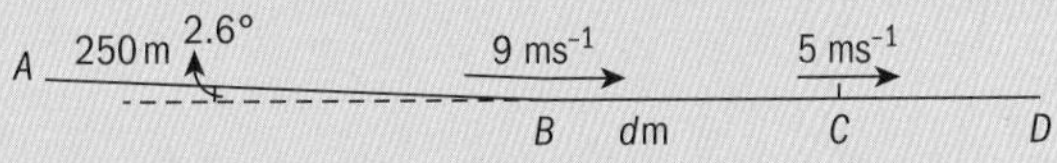

A cyclist and his machine have a total mass of 80 kg. The cyclist starts from rest at the top A of a straight path AB, and freewheels (moves without pedalling or braking) down the path to B. The path AB is inclined at 2.6° to the horizontal and is of length 250 m (see diagram).

i) Given that the cyclist passes through B with speed $9\,\text{ms}^{-1}$, find the gain in kinetic energy and the loss in potential energy of the cyclist and his machine. Hence find the work done against the resistance to motion of the cyclist and his machine.

The cyclist continues to freewheel along a horizontal straight path BD until he reaches the point C, where the distance BC is d m. His speed at C is $5\,\text{ms}^{-1}$. The resistance to motion is constant, and is the same on BD as on AB.

ii) Find the value of d.

The cyclist starts to pedal at C, generating 425 W of power.

iii) Find the acceleration of the cyclist immediately after passing through C.

Cambridge AS and A Level Mathematics 9709, Paper 4 Q5 May/June 2009

19.

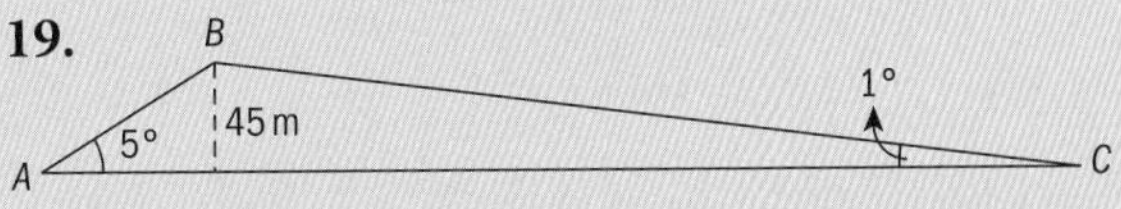

AB and BC are straight roads inclined at 5° to the horizontal and 1° to the horizontal respectively. A and C are at the same horizontal level and B is 45 m above the level of A and C (see diagram, which is not to scale). A car of mass 1200 kg travels from A to C passing through B.

i) For the motion from A to B, the speed of the car is constant and the work done against the resistance to motion is 360 kJ. Find the work done by the car's engine from A to B.

The resistance to motion is constant throughout the whole journey.

ii) For the motion from B to C the work done by the driving force is 1660 kJ. Given that the speed of the car at B is 15 ms^{-1}, show that its speed at C is 29.9 ms^{-1}, correct to 3 significant figures.

iii) The car's driving force immediately after leaving B is 1.5 times the driving force immediately before reaching C. Find, correct to 2 significant figures, the ratio of the power developed by the car's engine immediately after leaving B to the power developed immediately before reaching C.

Cambridge AS and A Level Mathematics 9709, Paper 41 Q6 October/November 2011

20. The velocity of a particle at a time t after it passes a point O is given by $v = 2t^3 - 5t^2 - 24t + 6$. Find the velocity at the time when the acceleration is zero.

21. The displacement of a particle at a time t is given by the equation $s = 3t(t - 1)(3 - 2t)$. Find expressions for the velocity and acceleration of the particle.

22. The displacement s of a particle from a point O is given by the equation $s = 2t^3 - 3t^2 - 12t + 7$. Find the acceleration at the instant when the velocity is zero.

23. The velocity v of a particle is given by the equation $v = \frac{t^3}{3} + \frac{1}{16t}$. Find the times at which the acceleration is zero and the velocity at that instant.

24. A train moves between two stations so that the distance s from station A is given by $s = 10t^3 - 6t^5$ km. The train first comes to rest at station B. Find

a) the time taken to travel between the two stations

b) the distance between the two stations

c) the maximum speed of the train between the two stations.

25. The velocity of a particle v at time t is given by $v = 3t^2 - 4t + 5$. Find an expression for s the displacement of the particle from O in terms of t, given that $s = 5$ when $t = 1$.

26. The acceleration of a particle moving along a straight line through a point O is given by the equation $a = t^3 - 6t$. When $t = 0$, the velocity of the particle is 6 and its displacement from O is 5. Find expressions for the velocity and displacement of the particle.

27. The velocity of a particle moving along a straight line is given by the equation $v = 3t + \frac{1}{t^2}$ ms^{-1}. Find the distance travelled by the particle between $t = 2$ and $t = 4$.

28. The velocity of a particle that moves in a straight line is given by the equation $v = 3\sqrt{4t+1}$. Find expressions for its acceleration a and its displacement s from a point O on the line given that $s = 4$ when $t = 0$.

29. The velocity of a particle moving along a straight line through a point O is given by the equation $v = 3t^2 - 6t$. Find the maximum velocity of the particle and its maximum displacement from O given that $s = -3$ when $t = 0$.

30. A vehicle is moving in a straight line. The velocity $v\,\text{m s}^{-1}$ at time t s after the vehicle starts is given by

$v = A(t - 0.05t^2)$ for $0 \le t \le 15$,

$v = \dfrac{B}{t^2}$ for $t \ge 15$,

where A and B are constants. The distance travelled by the vehicle between $t = 0$ and $t = 15$ is 225 m.

i) Find the value of A and show that $B = 3375$. [5]

ii) Find an expression in terms of t for the total distance travelled by the vehicle when $t \ge 15$. [3]

iii) Find the speed of the vehicle when it has travelled a total distance of 315 m. [3]

Cambridge International AS and A Level Mathematics 9709, Paper 41 Q7 May/June 2010

31. A particle P starts from a fixed point O at time $t = 0$, where t is in seconds, and moves with constant acceleration in a straight line. The initial velocity of P is $1.5\,\text{m s}^{-1}$ and its velocity when $t = 10$ is $3.5\,\text{m s}^{-1}$.

i) Find the displacement of P from O when $t = 10$. [2]

Another particle Q also starts from O when $t = 0$ and moves along the same straight line as P. The acceleration of Q at time t is $0.03t\,\text{m s}^{-2}$.

ii) Given that Q has the same velocity as P when $t = 10$, show that it also has the same displacement from O as P when $t = 10$. [5]

Cambridge International AS and A Level Mathematics 9709, Paper 41 Q4 October/November 2010

32. A particle P starts at the point O and travels in a straight line. At time t seconds after leaving O the velocity of P is $v\,\text{m s}^{-1}$, where $v = 0.75t^2 - 0.0625t^3$. Find

i) the positive value of t for which the acceleration is zero, [3]

ii) the distance travelled by P before it changes its direction of motion. [5]

Cambridge International AS and A Level Mathematics 9709, Paper 41 Q4 May/June 2012

33. A particle P starts from rest at a point O and moves in a straight line. P has acceleration $0.6t\,\text{m s}^{-2}$ at time t seconds after leaving O, until $t = 10$.

i) Find the velocity and displacement from O of P when $t = 10$. [5]

After $t = 10$, P has acceleration $-0.4t\,\text{m s}^{-2}$ until it comes to rest at a point A.

ii) Find the distance OA. [7]

Cambridge International AS and A Level Mathematics 9709, Paper 41 Q7 October/November 2013

Maths in real-life

Aerodynamics

The study of aerodynamics is very important in the design of vehicles such as bikes, aeroplanes and motor vehicles as well as in the design of many other structures.

Aerodynamics is the field of science and engineering that deals with the effects of air moving around objects. Aerodynamicists use physical laws, mathematical analysis, wind tunnels and computer simulations to predict what will happen in a given situation.

Wind tunnels are used to test how planes, bikes, cars and many other objects move through the air at different speeds and to predict the forces generated. This helps engineers to improve the design of anything affected by wind. For bikes and cars the aim is for the object to be pushed towards the ground, with no lift. For aeroplanes the aim is to generate lift so that the plane can fly.

Wind tunnels can be all different shapes and sizes. They can be whole buildings, using powerful fans to test life size objects, or small tunnels used to test small models.

Engineers can also use computers to solve problems in aerodynamics. The computer solves complex mathematical equations that are based on Newton's laws of motion.

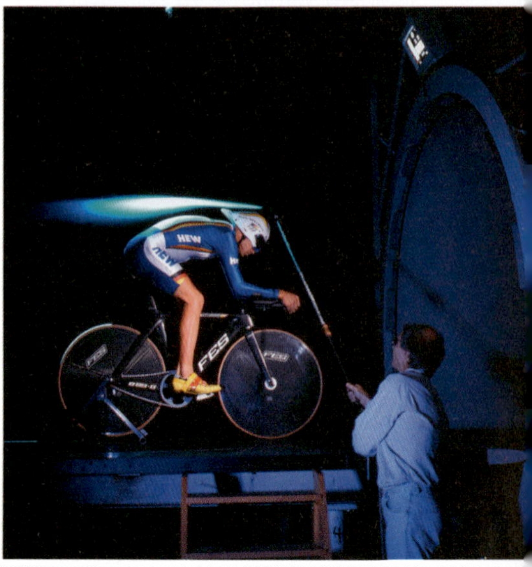

Aerodynamic testing is very important for cyclists. It is hard for a human being to improve their power output by 5%. This would require a lot of dedication and training. However improving their aerodynamics can make a 10 – 20% difference. Aero bars instead of an upright position is the most effective change, but an Aero helmet, skin suits and shoe covers all cut down on the drag a cyclist experiences. This means that less power is required to maintain the same speed.

Daniel Bernoulli, a Swiss scientist, developed a mathematical relationship between pressure and fluid flow in the 18th century. He was a leader in fluid mechanics, the study of liquids and gasses. The Bernoulli Principle is a mathematical explanation of why things can fly.

Engineers normally design wings, or airfoils, that are rounded on the top and flat on the bottom. This means that the pressure on the top of the wing is less than the pressure on the bottom of the wing, creating lift. Lift is a force that works upwards, opposing gravity.

The Wright brothers used wind tunnel testing to develop their first aircraft. The technology involved has developed significantly since then, and conducting a successful test is far more complicated than it may appear.

After a test is conducted, aerodynamicists need to make adjustments to the results in order to apply them to a full-scale aircraft and to suit the flight conditions. Judgement is very important at this stage. Any mistakes can have very serious consequences.

These adjustments can be made using theoretical and computational methods, or they can be made based on previous experiments and results.

▲ Red smoke is being used in the experiment photographed here to understand the turbulence created by an aircraft.

Exam-style paper A

55 Marks

1.

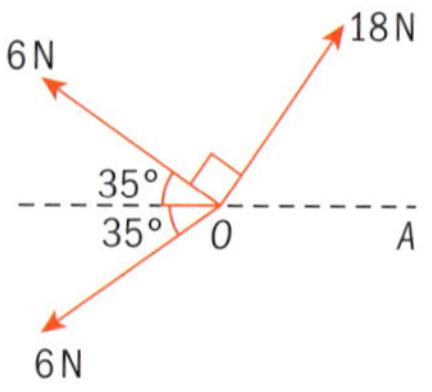

Three coplanar forces of magnitudes 6 N, 18 N and 6 N act at a point O in the directions shown in the diagram.

i) Find the component of the resultant of the three forces

a) in the direction of OA **b)** perpendicular to OA.

ii) Hence, find the magnitude and direction of the resultant of the three forces. **[6]**

2.

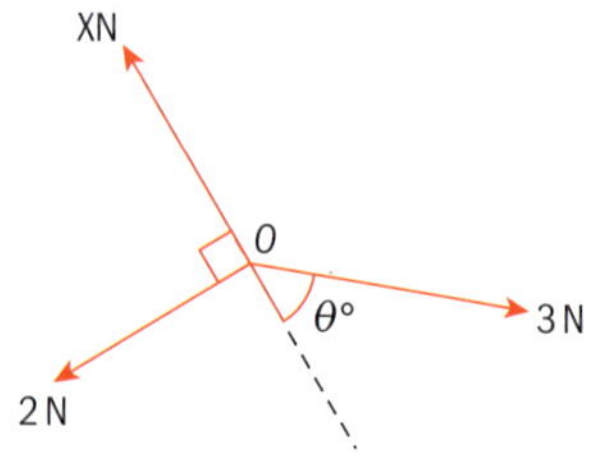

Three coplanar forces of magnitude X N, 3 N and 2 N are in equilibrium acting at a point O in the directions shown in the diagram. Find θ and X. **[4]**

3.

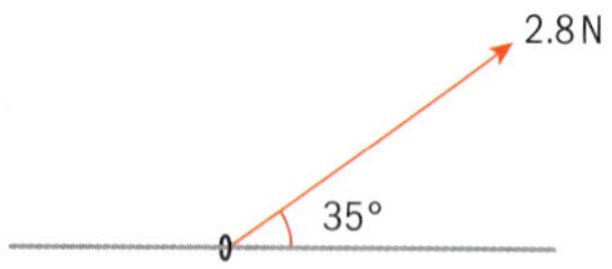

A shower curtain rail ring of mass 950 g is threaded on a rough curtain rail, which is fixed horizontally. Given that the ring is in equilibrium, acted on by a force of 2.8 N pulling upwards at 35° to the horizontal, find

i) the normal component of the contact force acting on the rail ring.

ii) The ring is in limiting equilibrium. Find the coefficient of friction between the ring and the rail. **[5]**

4. A tram moves along a straight road between stops A and B. The tram is at rest when it starts at stop A and again when it reaches stop B. The tram moves with constant acceleration for the first 40 s, with constant speed of 8 ms^{-1} for the next 560 s and then finally constant deceleration of 0.08 ms^{-2}.

i) Calculate the acceleration of the tram during the first 40 s. **[1]**

ii) Find the time taken for the tram to decelerate before it reaches stop B. **[1]**

iii) Sketch the velocity-time graph for the journey and calculate the distance from A to B. **[4]**

iv) Two people along the road record the tram's speed as 6 ms^{-1}, one while it is accelerating and one while it is decelerating. Calculate the distance between these two people. **[3]**

5.

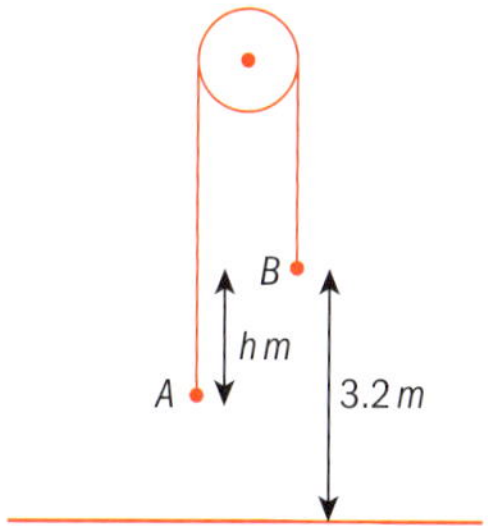

Particle A of mass 0.3 kg and particle B of mass 0.5 kg are attached at either end of a light inextensible string that passes over a smooth pulley. A is hanging h m below B which is being held at rest 3.2 m above the floor. Both parts of the string are vertical. B is released and both particles begin to move. When B reaches the floor, it remains at rest while A continues to move vertically upwards. A reaches its maximum height when it is 6.2 m above the floor. Calculate the velocity of the particles immediately before B reaches the floor and hence calculate the value of h. **[6]**

6. A racing car is designed to accelerate rapidly along a straight track. The car starts from rest at a point O and accelerates until it crosses the finishing line at a speed of $85\,\text{ms}^{-1}$. While the car is moving along the track t seconds after leaving O, its acceleration is $(6 + 0.4t)\,\text{ms}^{-2}$. Find

   i) the time taken by the car to reach the finishing line **[4]**

   ii) the distance travelled by the car along the track from O to the finishing line. **[3]**

7. A particle moves with constant acceleration $0.8\,\text{ms}^{-2}$ along a horizontal straight line ABC. The speed of the particle at C is $60\,\text{m s}^{-1}$ and the times taken from A to B and B to C are 30 s and 20 s respectively. Find

   i) the speed of the particle at A, **[2]**

   ii) the distance BC. **[3]**

8. A ball A is thrown vertically upwards with speed $16\,\text{ms}^{-1}$ from a point P. One second after the projection of A, a second ball B is also thrown vertically upwards from P with speed $16\,\text{ms}^{-1}$. Find

   i) the time that A has been moving when the balls collide, **[5]**

   ii) the height above P at which the balls collide. **[2]**

9. A force F is acting vertically upwards on a body of mass 10 kg. The body moves vertically from rest to a point A that is at a height 5 m above its starting point. The body has a speed of $7\,\text{ms}^{-1}$ at A.

   i) Find the work done by F. **[4]**

   It is now given that F is constant.

   ii) Find F. **[2]**

75 Marks

1.

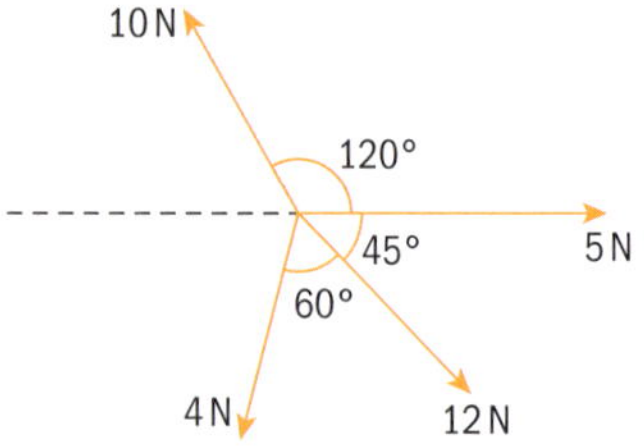

Three coplanar forces act at a point. The magnitude of the forces are 10 N, 5 N, 12 N and 4 N, and the directions in which the forces act are shown in the diagram. Find the magnitude and direction of the resultant of the four forces. **[6]**

2.

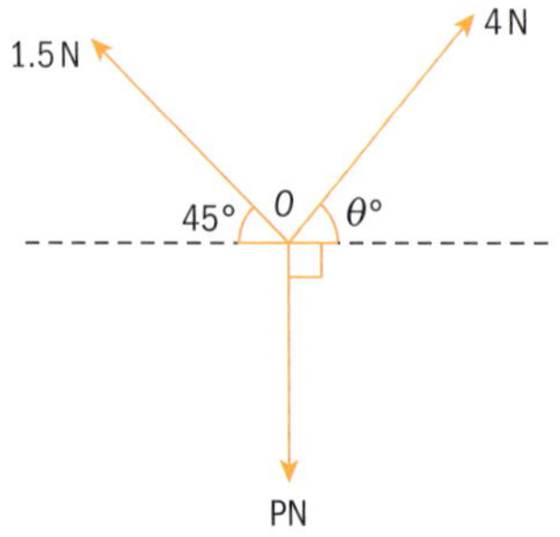

The three coplanar forces shown in the diagram act at a point O and are in equilibrium. Find the value of P and θ. **[6]**

3. A particle of mass 350 g rests on a rough plane inclined at an angle of $\theta°$ to the horizontal, where $\cos\theta = 0.96$. A force of magnitude 0.45 N, acting upwards on M at an angle $\theta°$ from a line of greatest slope of the plane, is just sufficient to prevent the particle from sliding down the plane. Find

   i) the normal component of the contact force on M **[2]**

   ii) the frictional component of the contact force on M **[3]**

   iii) the coefficient of friction between M and the plane. **[2]**

4. A car travels along a straight road from A to B, a distance of 5 km in 655 s. The car starts from rest at A and accelerates for T_1 s at 0.4 ms^{-1}. It reaches a speed $V\,\text{ms}^{-1}$ and then travels at constant speed for T_2 s. It decelerates for 40 s before coming to rest at B.

   i) Sketch the velocity-time graph for the motion of the car. **[1]**

   ii) Express T_1 and the final deceleration in terms of V. **[2]**

   iii) Express the total distance travelled in terms of V and show that $V^2 - 505\,V + 4000 = 0$. Hence find the value of V. **[5]**

5. 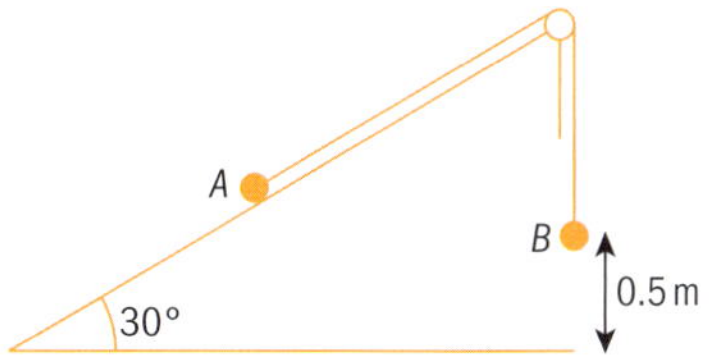

Particle A of mass 0.6 kg and B of mass 0.9 kg are connected by a light inextensible string that passes over a smooth pulley which is fixed at the top of a smooth, sloping, flat surface that is inclined at an angle of 30° to the horizontal. Particle B is released from rest and falls 0.5 m to the floor where it remains at rest. Particle A then continues to move up the slope.

i) Show that the magnitude of the acceleration of the particles is $2\,\text{ms}^{-2}$ and find the tension in the string. **[4]**

ii) Calculate the speed of the particles just before particle B reaches the floor. **[2]**

iii) Assuming that it does not reach the pulley, calculate the further distance that particle A travels up the slope. **[4]**

6. A particle P moves along a straight line through a point O. Its velocity at time t seconds is given by $v = 3.2 + t^{\frac{3}{2}}\ \text{ms}^{-1}$.

i) Find the acceleration of P at the time when its velocity is $11.2\,\text{ms}^{-1}$. **[4]**

ii) If the distance of P from O when the velocity is $11.2\,\text{ms}^{-1}$ is 27 m, find its distance from O when $t = 1$ second.

7. A car of mass 1200 kg travels up a line of greatest slope of a straight road inclined at 1° to the horizontal. The power of the car's engine is constant and equal to 25 kW and the resistance to the motion of the car is constant and equal to 750 N. The car passes through point A with speed $10\,\text{ms}^{-1}$.

i) Find the acceleration of the car at A. **[4]**

The car later passes through a point B with speed $15\,\text{ms}^{-1}$.
The car takes 28 seconds to travel from A to B.

ii) Find the distance AB. **[6]**

8. The resistance to the motion of a lorry of mass 5 tonnes is kv, where $v\,\text{ms}^{-1}$ is the lorry's speed. With the engine working at 66 kW, the lorry can attain a greatest steady speed of $12\,\text{ms}^{-1}$ up a straight road which is inclined at $\sin^{-1}\frac{1}{20}$ to the horizontal.

i) Show that $k = 250$. **[4]**

ii) Find the greatest steady speed that lorry can go down this slope with the engine still working at 66 kW. **[4]**

9. A block of mass 25 kg is dragged 30 m up a slope inclined at 5° to the horizontal by a rope inclined at 20° to the slope. The tension in the rope is 100 N and the resistance to the motion of the block is 70 N. The block is initially at rest. Calculate

i) the work done by the tension in the rope, **[2]**

ii) the change in the potential energy of the block, **[2]**

iii) the speed of the block after it has moved 30 m up the slope. **[4]**

Answers

The answers given here are concise. However, when answering exam-style questions, you should show as many steps in your working as possible.

Cambridge International Examinations bears no responsibility for the example answers to questions taken from its past paper questions which are contained in this publication.

1 Straight line motion and graphs

Skills Check page 2

1. 1200 **a)** 0.375, **a)** 0.375,

2. a) 0.375, **b)** −0.5

3. a) 6 s **b)** 40 s **c)** 0.2 s

4. a) 144 m **b)** 4 m **c)** 15 m

5. a) 3 ms^{-1} **b)** 1.5 ms^{-1} **c)** 0.2 ms^{-1}

6. a) 100 ms^{-1} **b)** 6 ms^{-1}

7. a) 3 ms^{-2} **b)** −2 ms^{-2}

8. 3 s

Exercise 1.1 page 5

1. a)

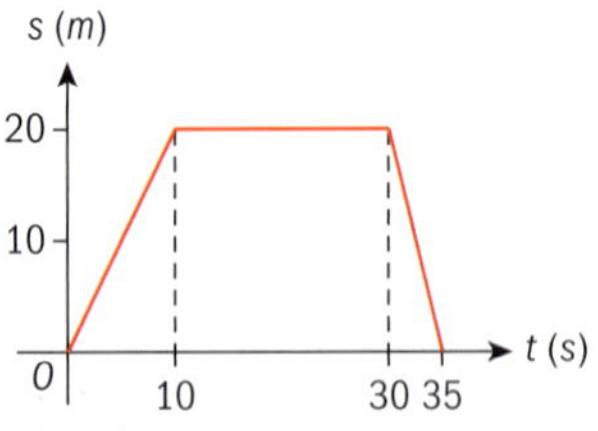

b) −4 ms^{-1}

2. a)

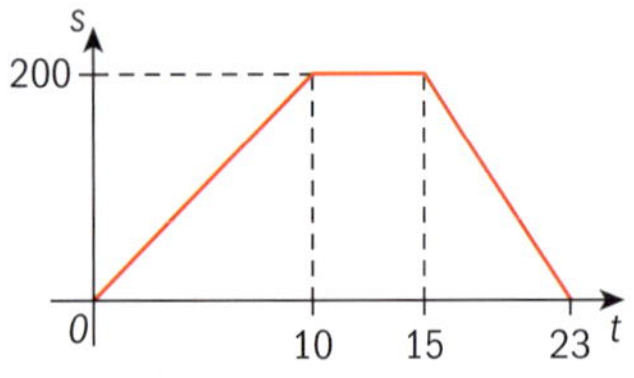

b) −25 ms^{-1}

3. a)

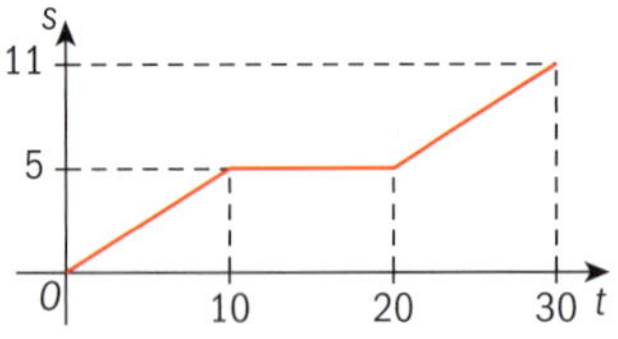

b) 11 m

4. a) 1.5 ms^{-1} **b)** 0 **c)** 0.5 ms^{-1}

5. a) 1.6 ms^{-1} **b)** 0 **c)** 0.8 ms^{-1}

6. a) −1.125 ms^{-1}

b) 0 **c)** −0.75 ms^{-1}

7. a) 320 m **b)** −16 ms^{-1}

8. a) 0.667 s **b)** 2 s

9. a) −8 ms^{-1} **b)** 72 m

10. a) 1.6 ms^{-1} **b)** −8 m **c)** 12 m

d) −20 m **e)** −2 ms^{-1}

Exercise 1.2 page 12

1.

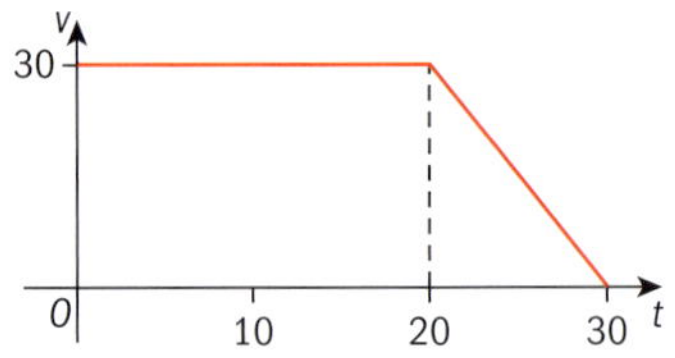

2.

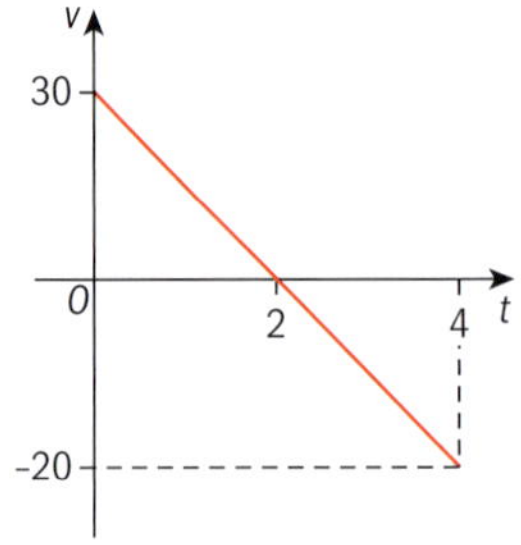

3.

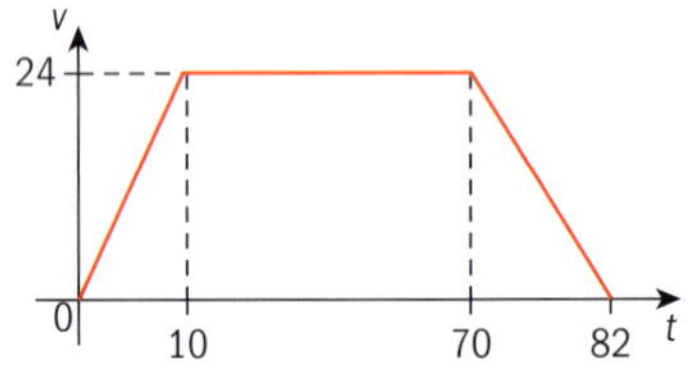

4. a) 0.167 ms^{-2} and −0.1 ms^{-2}

b) 450 m **c)** 3.46 ms^{-1}

5. a) 1.6 ms^{-2} **b)** 240 m **c)** 660 m

6. a) 15 ms^{-1} **b)** 0.75 ms^{-2} **c)** 375 m

7. a) 30 s **b)** 750 m **c)** 4.17 ms^{-1}

8. a) 2.5 ms^{-2}, −2.5 ms^{-2}, −2.5 ms^{-2}, 2.5 ms^{-2}.

b) 25 m **c)** 35 m **d)** 4th floor

9. a) 24 ms^{-1} **b)** 3.43 ms^{-2}

c) −3 ms^{-2} **d)** 1.75 s and 23 s

10. a) 10 ms^{-1}

b) −2 ms^{-2} and −0.625 ms^{-2}

c) 10.6 ms^{-1} **d)** 13 s

11. a) 5 ms^{-1} or 1 ms^{-1} **b)** 2 s or 2.8 s

12. a) 4 ms^{-1} **b)** 4 s

13. a) 45 m **b)** 5 s **c)** −30 ms^{-1}

d) 15 ms^{-1} **e)** 11.25 m **f)** 8 s

Summary exercise 1 page 16

1. a) 0.4 m

b)

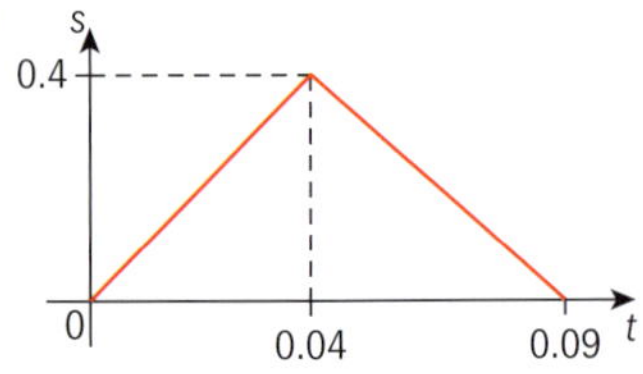

2. a)

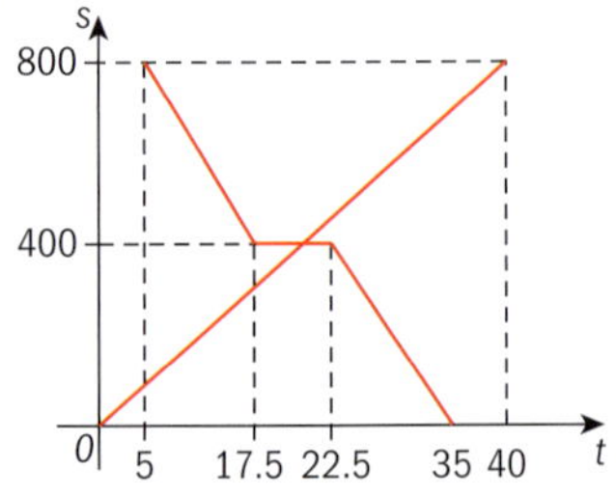

b) 400 m, 20 s

c) the train moving from B to A (35 s)

3. a) 240 m **b)** 13.3 ms^{-1}

c) −30 ms^{-2} **d)** 180 m

4. a) 9.83 s **b)** 3.6 ms^{-2}

5. i) 3600 s

ii)

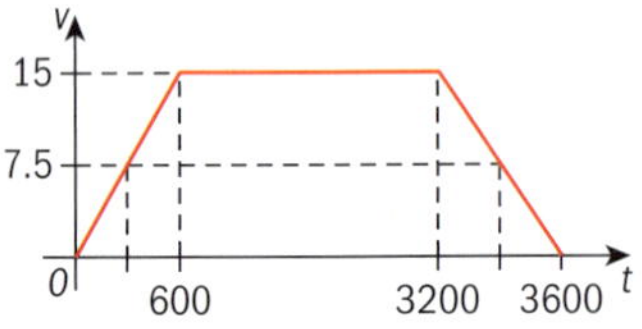

46, 500 m

iii) 300 s or 3400 s

6. i) 0.09 ms^{-2} **ii)** 1.08 m **iii)** 0.72 ms^{-1}

2 Constant acceleration formulae

Skills Check page 18

1. a) 14 **b)** 18 **c)** 2

2. a) $u = 4$ **b)** $a = \frac{8}{3} = 2\frac{2}{3}$ **c)** $t = 1\frac{1}{2}$

3. a) 2, 6 **b)** −0.281 or 1.78

c) 0.869 or −1.54

Exercise 2.1 page 23

1. 128 m
2. 40 m
3. 15 ms^{-1}
4. 5 ms^{-1}
5. 9.8 ms^{-2}
6. 20 s
7. 2 s
8. 12 m
9. 4 ms^{-1}
10. 23 ms^{-1}
11. a) 500 m b) 20 ms^{-1}
12. a) 31 m b) 5 s
13. 4.25 ms^{-1}
14. 37 m
15. 17.1 m
16. i) $ad_1 = 12$, $ad_2 = 7.5$ ii) $d_1 = 1.6d_2$

Exercise 2.2 page 25

1. 6 ms^{-1}
2. 3.10 s
3. 16.2 m
4. 5 ms^{-1}, 30 m
5. 31.25 m, 5 s
6. a) 15.6 ms^{-1} b) 2.76 s
7. a) 2.45 s b) 24.5 ms^{-1}
8. a) 0.8 s b) 4.2 m c) 9.17 ms^{-1}
9. 30 m
10. a) i) 1 s ii) 5 s b) 4 s
11. a) 20 ms^{-1} b) 20 m c) 2.83 s
12. a) 5.74 ms^{-1} b) 4.05 m c) 1.36 s
13. a) Proof b) 6 ms^{-1}
    c) 1 ms^{-1} upwards, 1 ms^{-1} downwards
14. i) 20 ms^{-1} ii) 3 s iii) 35 m

Summary Exercise 2 page 28

1. a) 2 s b) 2 s
2. a) 320 ms^{-1} b) 1600 m
3. a) 11 m b) 10.1 s
4. 7.22 s, 47.2 ms^{-1}
5. 7.81 s
6. 25.8 s, 200 m
7. 68.3 s, 46.8 ms^{-1}
8. i) 0.5 ms^{-2} ii) $\alpha = 2.9$
9. i) 9 ms^{-1} ii) 3 s
10. i) 1.5 m ii) 1.05 m

3 Forces and resultants

Skills Check page 30

1. $\overrightarrow{OC} = 3\mathbf{a}$, $\overrightarrow{OC} = -3\mathbf{a}$
2. 53.1°

Exercise 3.1 page 34

1. a) 21.7 N, 10.9° b) 20.3 N, 24.7°
   c) 13.5 N, 59.2°
2. a) 493.17 N, 12.3° b) 250 N, 36.9°
   c) 851 N, 20.4° d) 164 N, 53.1°
3. 363 N, 8.7°
4. 84.5°, 45.3°
5. 66.6 N
6. 47.1°

Exercise 3.2 page 36

1. a) 14.1 N b) 14.1 N
   c) $(14.1\mathbf{i} + 14.1\mathbf{j})$N
2. a) −13 N b) −7.5 N
   c) $(-13\mathbf{i} - 7.5\mathbf{j})$ N
3. a) 12.5 N b) −21.65 N
   c) $(12.5\mathbf{i} - 21.65\mathbf{j})$N
4. a) −7.85 N b) 3.66 N
   c) $(-7.87\mathbf{i} + 3.66\mathbf{j})$ N
5. a) −24.5 N b) −24.5 N
   c) $(-24.5\mathbf{i} - 24.5\mathbf{j})$ N
6. a) −10.6 N b) 19.7 N
   c) $(-10.6\mathbf{i} + 19.7\mathbf{j})$ N

Summary exercise 3 page 36

1. i) 5.50 N, −8.49 N, −2.83 N;
   3.85 N, 8.49 N, −4.90 N;
   $(5.50\mathbf{i} + 3.85\mathbf{j})$N, $(-8.49\mathbf{i} + 8.49\mathbf{j})$ N,
   $(-2.83\mathbf{i} - 4.90\mathbf{j})$ N
   ii) $x = (6 + 6\sqrt{2})$N; $y = 6\sqrt{2}$ N;
   $(6 + 6\sqrt{2})\mathbf{i} + 6\sqrt{2}\mathbf{j}$
   iii) $x = (-3 + 4\sqrt{2})$N; $y = 4\sqrt{2}$ N;
   $(-3 + 4\sqrt{2})\mathbf{i} + 4\sqrt{2}\mathbf{j}$
   iv) $x = 2$N; $y = 2\sqrt{3}$ N; $2\mathbf{i} + 2\sqrt{3}\mathbf{j}$
   v) $x = 3.10$ N; $y = 1.64$ N; $3.10\mathbf{i} + 1.64\mathbf{j}$
   vi) $x = -6.02$ N; $y = -3.64$ N;
   $-6.02\mathbf{i} - 3.64\mathbf{j}$
   vii) $x = 4.94$ N; $y = 2.98$ N; $4.94\mathbf{i} + 2.98\mathbf{j}$
2. a) $x = 3.40$ N; $y = 11.90$ N; $R = 12.40$ N
   b) $x = -4\sqrt{2}$ N; $y = 2.19$N; $R = 6.07$ N
   c) $x = -4.93$ N; $y = 3.63$ N; $R = 6.12$ N
   d) $x = -9 + 6\sqrt{2}$ N; $y = -1.10$N;
   $R = 1.22$N
   e) $x = -2 + 3\sqrt{3}$ N; $y = 5$N; $R = 5.93$N
   f) $x = 2.32$N; $y = -9 - \sqrt{2}$ N;
   $R = 10.67$ N
3. a) $R = 4.91$ N; $\theta = 6.12°$
   b) $R = 4.26$ N; $\theta = 35.1°$
   c) $R = 7.46$ N; $\theta = 152.7°$
   d) $R = 10.43$ N; $\theta = 205.5°$ (25.5° *under the negative x-axis*)
4. $R = 573.02$ N
5. $R = 73.1$ N
6. $X = 173$ N
7. 3 N
8. 140°
9. $\sqrt{30}$
10. $R = 53.3$ N
    $\theta = 19.0°$ to the $\mathbf{i}$ direction.
11. $Q = 8.91$
12. i) a) 8.74 N b) 11.5 N
    ii) 14.4 N, 52.7° anticlockwise from $\mathbf{i}$ direction.
13. $R = 6.94$ N
    $\theta = 58.6°$ below the positive x-axis.
14. 500 N, 36.9°
15. i) 67.4 ii) 9 N

Review exercise A page 41

1. a)
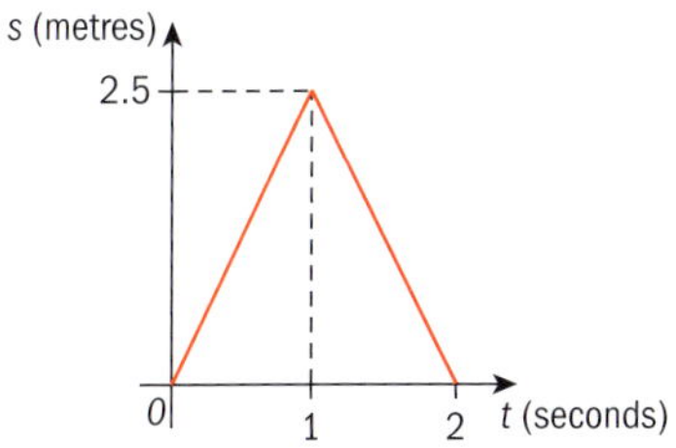

   b) 2.5 ms^{-1}
2. a)
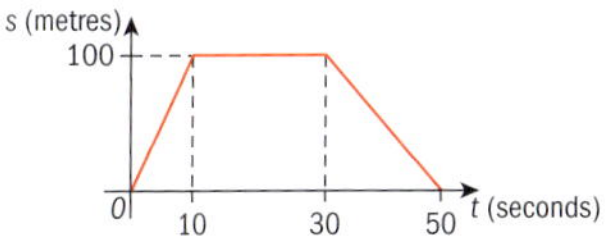

   b) 10 ms^{-1} and −5 ms^{-1}

3. a) 210 m, 8100 m, 900 m, 90 m
9.3 km 15.5 ms⁻¹

b) 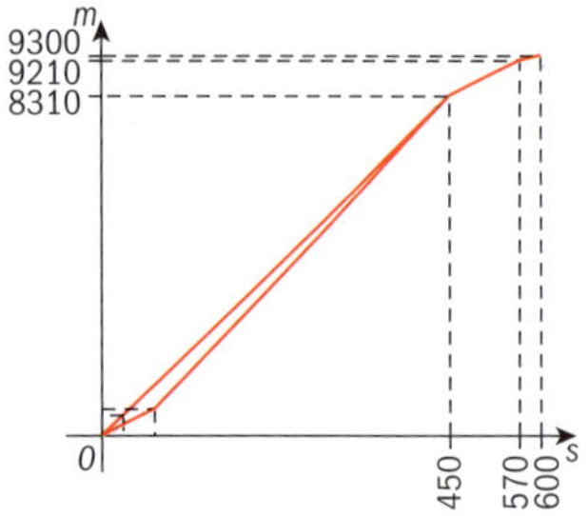

4. a) 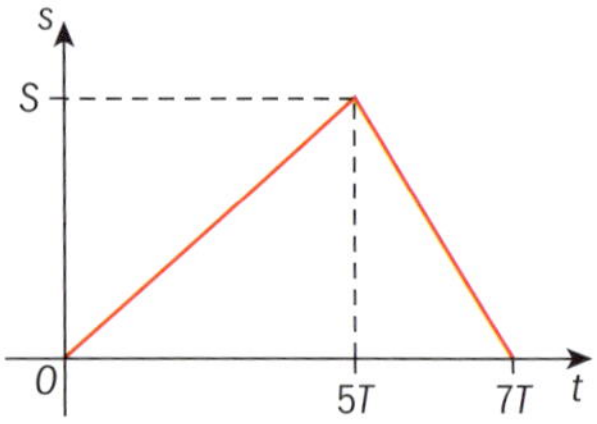

b) −75 ms⁻¹ c) 750 m

5. a) 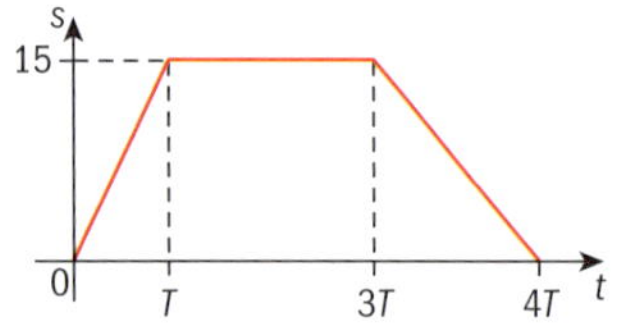

b) 0.75 c) 3 s

6. a) 0.2 ms⁻² b) 0.1 ms⁻²
c) −0.8 ms⁻² d) 990 m

7. a) 25 s
b) −0.5 ms⁻² and 412.5 m

8. a) 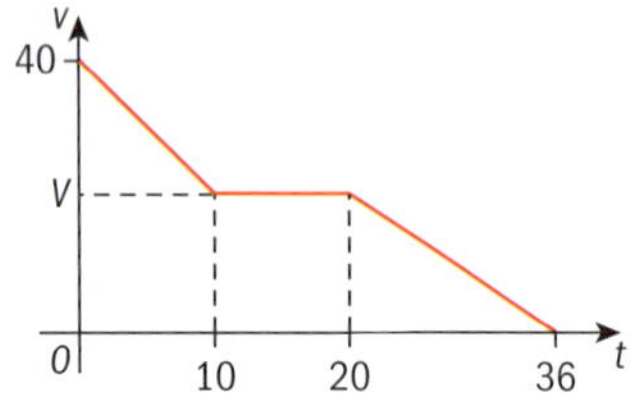

b) 20 ms⁻¹ c) 18.3 ms⁻¹

9. a) 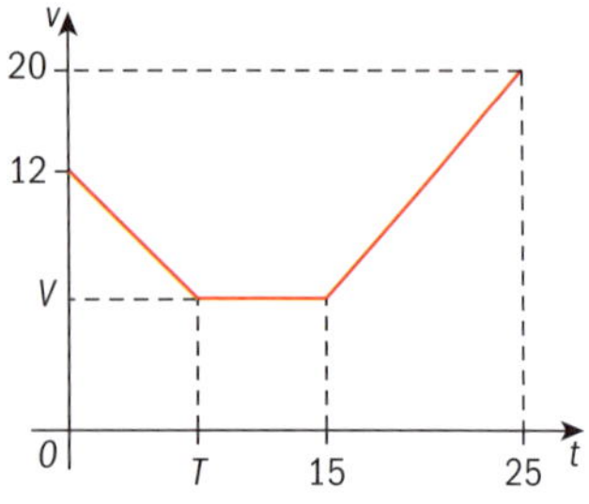

b) 2 c) 7.8

10. a) 123 ms⁻¹ b) 18.9 ms⁻²

11. i) $V = 20$ ms⁻¹ ii) 40 ms⁻¹
iii) 80 m

12. i) 3600 s
ii) 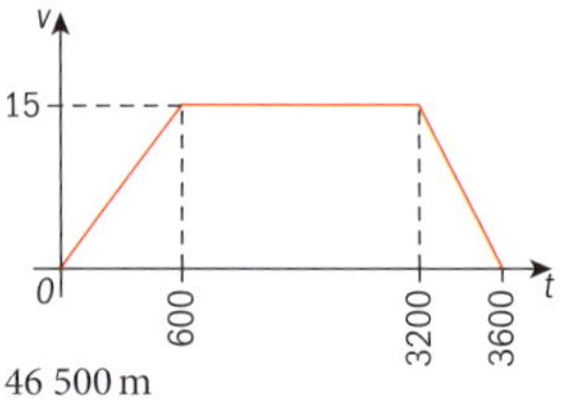

46 500 m

iii) $t = 300$ s or $t = 3400$ s.

13. 3 ms⁻¹, 150 m

14. a) 4 ms⁻¹ b) 7 s

15. a) 9.6 ms⁻¹ b) 13.5 s

16. a) 0.2 s b) 1.2 m c) 4.90 ms⁻¹

17. a) 0.2, 0.92 b) 0.72 s c) 1.176 m

18. 35 m

19. i) 2.5 ms⁻¹ ii) 14.5

20. 6.8 m

21. i) 30 ms⁻¹ ii) 1 s iii) 35 ms⁻¹

22. R = 97.9 N (1d.p.)

23. R = 62.2 N (1d.p.)

24. R = 212.6 N (1d.p.) $\theta = 13.6°$

25. R = 478.2 N (1d.p.) $\theta = 16.8°$

26. $20.34\mathbf{i} + 44.02\mathbf{j}$

27. $-6.57\mathbf{i} + 30.03\mathbf{j}$

28. $-20.93\mathbf{i} - 29.76\mathbf{j}$

29. R = 5.37 N $\theta = 33.9°$

4 Newton's laws

Skills Check page 48

1. a) $s = 24$ b) $u = 12$ c) $s = 2.5$

2. horizontal = 10.4 N, vertical = 6 N

Exercise 4.1 page 52

1. a) Slides backwards.
b) Stays where it is.
c) Slides forwards.

2. $a = 3$ ms⁻²

3. $m = 15$ kg

4. $F = 48$ N

5. $t = 4$ s

6. $v = 12$ ms⁻¹

7. $F = 2.0$ N

8. $F = 156$ N

9. $s = 2.81$ m

10. a) $T = 20\,500$ N b) $T = 6980$ N
c) $m = 770$ kg

Exercise 4.2 page 54

1. a) i) 1280 N ii) 14 600 N
b) i) 8430 N ii) 12 000 N
c) i) 14 400 N ii) 3060 N

2. 12°

3. Parallel: 13 860 N, Perpendicular: 16 520 N;
a) Parallel: 12 470 N, Perpendicular: 14 860 N
Parallel: 11 280 N, Perpendicular: 13 440 N
Parallel: 10 330 N, Perpendicular: 12 310 N
b) 4210 N

4. 92 kg

Exercise 4.3 page 55

1. a) $P = 30, Q = 60$ b) $P = 24, Q = 18$

2. 7**i** is the resultant force, so accelerating in positive x-direction.

3. 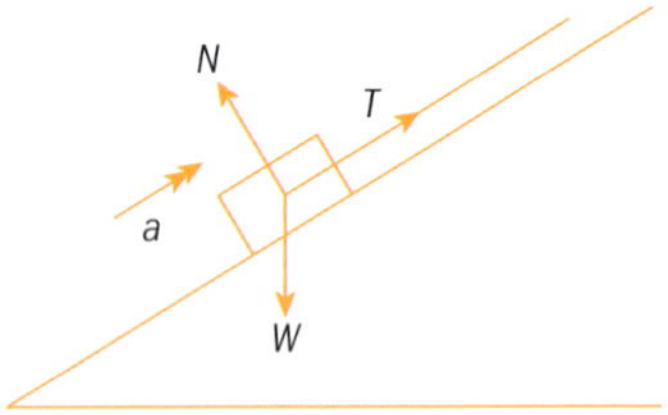

4. 23.4 N

5. $T = 4 + 1.5g$

6. $\theta = 8.74°$, $a = 0.0925$ ms⁻²

7. $a = 2.65$ ms⁻², $s = 0.477$ m

8. $L = 932$ N, $R = 98.0$ N

Exercise 4.4 page 58

1. $\frac{2}{3}g$

2. $\frac{g}{12}$; $1\frac{5}{6}g$

3. 0.904 ms⁻²

4. $F = 40\,200$ N

Summary exercise 4 page 60

1. $T = 4.55$ N, $R = 9.1$ N

2. a) i) 676 N ii) 8480 N
b) i) 611 N ii) 7660 N

3. a) $T = 2$ N b) $F = 1.2$ N

4. a) $a = 0.4\,ms^{-2}$ b) $s = 203\,m$

5. i) $a = 1\,ms^{-2}$

ii) a) $s_P = 3\,m, s_Q = 7\,m$

b) $v = 2\,ms^{-1}$

iii) Proof

5 Equilibrium

Skills Check page 62

1. 19.4 cm
2. 3.8 cm
3. 101.54°

Exercise 5.1 page 67

1. a) 20.7 N, 137° b) 10.3 N, 131.4°

c) X = 3.9 N, Y = 0.7 N

d) 6.4 N, 128.7° e) 8.1 N, 111.7°

f) 6.87 N, 71.4° g) 49.7 N, 5.3°

2. a) $P = 5.57\,N, Q = 4.47\,N$

b) $P = 9.2\,N, Q = 7.96\,N$

c) $P = 21\,N, Q = -9\,N$

3. $X = 56.2\,N, R = 80.3\,N$
4. $X = 46.9\,N, R = 116.3\,N$
5. $R = 22.6\,N, T = 39.2\,N, m = 4.62\,kg$
6. $R = 22.6\,N, T = 39.2\,N, m = 4.1\,kg$

Summary exercise 5 page 69

1. a) a = −5 b = −6

b) c = 9 d = 5

c) e = 1 f = −4

d) g = 5 $h = (7 - \sqrt{3})$

2. a) A = 74.31 N b) $A = 5\sqrt{39}\,N$

c) A = 12.21N d) A = 19.15 N

e) A = 22.52 N f) A = 25.88 N

3. a) A = 2.88 N; B = 11.60 N

b) A = −5.73 N; B = 28.42 N

c) A = 14.15 N; B = 1.79 N

d) A = 37.94 N; B = 41.84 N

e) A = −89.30 N; B = −124.15 N

4. X = 7.34 N; R = 67.60 N
5. Y = 1478.63 N; R = 3132.02 N
6. R = 45.71 N; T = 32 N; $m = 5.58\,kg$
7. R = 87.98 N; T = 103.92 N; $m = 13.16\,kg$
8. Proof
9. Proof

10. 2.52 N

11. A = 3

12. i) T = 3 N. ii) F = **2.4 N.**

iii) $W = 1.4$

6 Friction

Skills Check page 72

1. a) $\cos A = \frac{15}{17}$; angle CAB = 28.1°2.

b) $\cos Q = \frac{3}{5}$; angle PQR = 53.1°

c) Longest side XZ = 25 cm; angle ZXY = 53.1°

2. a) $a = \frac{v-u}{t}$,

b) $a = \frac{2(s-ut)}{t^2}$,

c) $t = \frac{g^3}{p^4}$

3. a) $v = \pm\,6.24$ (2 d.p.)

b) $s = 50.5$

c) $p = \pm 0.62$ (2 d.p.)

Exercise 6.1 page 76

1. No motion will occur.
2. No motion will occur.
3. Motion will take place.
4. Motion will take place.
5. No motion will occur.
6. Motion will occur.

Exercise 6.2 page 79

1. a) Motion will take place down the plane; $F_{max} = 28.2\,N$

b) Motion will take place down the plane; $F_{max} = 31.8\,N$

c) Motion will take place up the plane; $F_{max} = 155.9\,N$

2. To move up the plane, force must exceed 1183.3 N
3. 418.8 N
4. 0.93
5. 0.36
6. 0.34

Summary exercise 6 page 81

1. $4.42\,ms^{-2}$
2. 0.77
3. 4.9 m
4. $30\,ms^{-1}$
5. a) $F_{max} > 20\,N$, no motion

b) $F_{max} > 11\sqrt{3}\,N$, no motion

c) $F_{max} < 35\sqrt{6}\,N$, motion takes place $a = 10.4\,ms^{-2}$

6. a) X = 94.4 N b) X = 9.30 N

c) X = 97.0 N d) X = 265 N

7. a) Since 306 N > 260 N, no motion takes place.

b) Since 376 N > 200 N, no motion takes place.

c) Since 900 N > 630 N, motion will take place.

d) Since 283 N > 173 N, motion will take place.

e) Since 508 N > 47 N, no motion takes place.

8. $\mu = 0.7$
9. $a = 0.87\,ms^{-2}$
10. $\mu = 0.19$
11. $\mu = 0.74$
12. a) $\theta = 12.3°$

b) $\mu = 0.218$

13. $\mu = 0.575$

Review exercise B page 84

1. $R = 0.461\,N$
2. $R = 250\,N$

$s = 3750\,m$

3. 71.4 N
4. $0.624\,ms^{-2}$
5. $1.19\,ms^{-2}$
6. $R = 58.5\,N$
7. $3.5\,ms^{-2}$
8. $T = 205\,N, R = 385\,N, \theta = 20°$
9. $a = 3\,ms^{-2}, T = 21$
10. $a = 1.67\,ms^{-2}$
11. i) $a = 2\,ms^{-2}$

ii) $T = 3.6\,N$

iii) $m = 0.3\,kg$

iv) 0.792 m.

12. i) 3.84 N; $1.6\,ms^{-2}$

ii) 1.5 s

13. i) $F = 102.8\,N$

ii) $s = 122.5\,m$

14. i) $T = 4.2$ N

ii) $t = 0.6$ s

15. X = 10.58 N
$\theta = 139.11$

16. X = 10.72 N (2 d.p.)
Y = 9.00 N (2 d.p.)

17. X = 27.53 N (2 d.p.)
$\theta = 86.26°$ (2 d.p.)

18. $F = 57.15$ N (2 d.p.)
$\theta = 81.46$ (2 d.p.)

19. $a = 3$, $b = -5$

20. $F = 140$ N; $F_{max} > 130$ N $\Rightarrow$ no motion takes place.

21. $F = 350$ N; $F_{max} > 400$ N $\Rightarrow$ no motion takes place.

22. $F = 1706.25$ N; 1900 N > 1856.25 N $\Rightarrow$ no motion takes place.

23. $F = 1777.73$ N; $F_{max} > 1767.77$ N $\Rightarrow$ no motion takes place.

24. $F = 1188.69$ N; 1488.69 > 975 N $\Rightarrow$ no motion takes place.

25. $F = 2582.63$ N; 3832.63 > 3586.71 N $\Rightarrow$ no motion takes place.

26. $\mu = 0.27$

27. $a = 0.33$ ms^{-2}

28. i) $F = 3.80$ N; $\tan\theta = 0.5$

ii) 7.60 ms^{-2}, 26.6°

7 Work and energy

Skills Check page 90

1. a) 25 N **b)** 91.9 N **c)** 51.4 N

2. a) 730 N **b)** 4590 N

Exercise 7.1 page 93

1. 800 J

2. 12 800 J

3. 2700 J

4. 1264 J

5. 3900 J

6. 130 J

7. 71 J

8. 8 N

9. 20.3 N

10. a) 550 m, 66 000 J **b)** 607 m, 68 500 J

Exercise 7.2 page 95

1. a) 40 J **b)** 9 J **c)** 60 000 J **d)** 20 J

2. a) 7700 J **b)** 125 J

3. a) 12.5 J **b)** 43 200 J

4. 5.42 ms^{-1}

5. a) 0.2 J **b)** 3.58 ms^{-1}

6. a) 1500 J **b)** 1.58 ms^{-1}

Exercise 7.4 page 97

1. a) 680 J **b)** 187 000 J **c)** 66.5 ms^{-1}, assuming no air resistance

2. a) 7.2 m **b)** 18.72 m **c)** 12 ms^{-1}

3. a) 700 kJ **b)** 699 kJ

4. 17.3 ms^{-1}

5. 3.2 m

6. a) 25.6 J **b)** 12.8 m **c)** 1.5 m

7. 11.1 ms^{-1}

8. 7.2 m

9. 3 m

10. a) 12.0 ms^{-1} **b)** 12.0 ms^{-1} **c)** 15 ms^{-1} **d)** 11.25 m

11. 5.33 m

12. 44.1 m

Exercise 7.5 page 101

1. 4.5 N

2. 80 500 J, 1610 N

3. 588 880 J

4. 1875 N

5. 175 600 N

6. a) 2570 J **b)** 9.26 ms^{-1}

7. 5.37 ms^{-1}

8. a) 9.49 ms^{-1} **b)** 910 J, 75.8 N

9. a) 60 J **b)** 5.39 ms^{-1}

10. 0.1 ms^{-1}

Summary Exercise 7 page 103

1. i) 591 J **ii)** 414 J **iii)** 177 J

2. 13.5 ms^{-1}

3. 2820 J

4. a) i) 45 J **ii)** 5000 J **b)** 32.8°

5. 26.7 ms^{-1}

6. 8270 J

7. 97.9 N

8. 114 m

9. i) 7100 kJ **ii)** 24 m

8 Power

Skills Check page 106

1. a) 0.8 ms^{-2} **b)** 1.07 ms^{-2}

2. a) 6000 J **b)** 480 kJ **c)** 500 N

Exercise 8.1 page 109

1. 1800 W

2. 1600 W

3. 180 W

4. 21 W

5. a) 6.67 ms^{-1} **b)** 11.3 ms^{-1} **c)** 15.3 ms^{-1}

6. 778 N

7. 540 W

8. 625

9. a) 10.4 ms^{-1} **b)** 4.51 ms^{-1} **c)** 15.5 ms^{-1}

10. 6370 N

11. a) 6250 kW **b)** 13.9 ms^{-1}

12. a) 70 900 W **b)** 98.7 m **c)** 11.0 s

13. i) 25 000 J **ii)** 16.4 s

Exercise 8.2 page 112

1. 0.214 ms^{-2}

2. a) 4.6875 ms^{-2} **b)** 0.9375 ms^{-2} **c)** 48 ms^{-1}

3. a) 1.7 ms^{-2} **b)** 3.48 ms^{-1} **c)** 26.7 ms^{-1}

4. a) 1800 N **b)** 0.0225 ms^{-2}

5. a) 11 200 W **b)** 0.0794 ms^{-2}

6. a) 1250 N **b)** 2.17 **c)** 61 300 W

7. a) Proof **b)** 12.5 ms^{-1}

8. a) 1.38° **b)** 5 ms^{-1}

9. 16 000, 1500

10. a) 31.25 **b)** 28.4 ms^{-1} **c)** 56.3 ms^{-1}

11. i) $R = 600$ N **ii)** 0.25 ms^{-2}

12. 40 ms^{-1}

Summary Exercise 8 p

1. 0.845 ms^{-2}
2. a) 16 b) 320 N
3. a) 500 N b) 22.4 ms
4. i) 0.025 ms^{-2} ii) 5
5. $P = 21.9$
6. 165 000, 3660
7. a) 17.28 b) 2220 kg
8. a) Proof b) 30, 18
9. i) 0.4 ms^{-2} ii) Proof
10. i) 0.15 ms^{-2} ii) Proof

9 Variable forc

Skills Check page 11

1. a) $2x - 2$ b) $-\frac{1}{x^2} + $

   c) $\frac{1}{2\sqrt{x}}$
2. Minimum: (0, 0); Maxim
3. a) $x^3 + 2x^2 + c$ b) -

   c) $\frac{2}{3}x^{\frac{3}{2}} + c$ d)

   e) $\frac{1}{3}(2x-1)^{\frac{3}{2}} + c$
4. $-\frac{1}{6}$
5. 17

Exercise 9.1 page 12

1. $t = 1$
2. $v = 6\frac{1}{3}$
3. $a = 3t^2 - 12t + 9, v = 4$
4. $v = 6t^2 + 2t - 1, a = 12t + 2$
5. $v = 8t^3 - 27, a = 24t^2, a = 54$
6. $v = 6t^2 - 6t, a = 12t - 6,$
   $s = -73$ and $s = -72$
7. $a = 2t - \frac{1}{4t^2}, t = \frac{1}{2}, v = \frac{3}{4}$
8. a) $t = 9$ s b) $t = 4$ s, $a = -\frac{3}{32}$ ms^{-2}
9. a) $t = 1$ hour b) $s = 2$ km

   c) $v = 3\frac{5}{9}$ kmh^{-1}
10. a) $v = 6t^2 - 18t + 12$

    b) $t = 1$ and 2

    c) $s = 1$ and 0 d) $s = 11$

25

$^2 - \frac{1}{4}t^4 - 8t - 2$

$\frac{1}{3}, s = 7\frac{1}{3}$

$- 1)^{\frac{5}{2}}$

9 page 127

) 25 m

) 24 s (and 0 s)

) 7.25 s

i) 16.9 ms^{-1}

v) 144 s

-2 ms^{-1} and 0

m

page 129

b) 40 m

b) 6.73 ms^{-1}

4. 5.95 ms^{-1}
5. 3700 N
6. 31.1 N
7. i) 7 ms^{-1}

   ii) 21.1° or 0.368°

   iii) 2.8 J
8. i) 32 000 J ii) 125 J iii) 1250 W
9. i) 4.99×10^6 J ii) 1500 m
10. 200 W
11. 300 W
12. a) 50 ms^{-1} b) 2 ms^{-2}
13. 19 600 W
14. 1000 N
15. 20 000
16. i) 25 000 J

    ii) 20 s
17. i) Proof

    ii) 2 kW
18. i) 3240 J; 9070 J; 5830 J

    ii) 96.0

    iii) 0.771 ms^{-2}
19. i) 900 000 J

    ii) Proof

    iii) 0.75
20. $v = -57$ ms^{-1}
21. $v = \frac{ds}{dt} = -18t^2 + 30t - 9$

    $a = \frac{dv}{dt} = -36t + 30$
22. $a = 18$ ms^{-2}
23. $t = \frac{1}{2}$

    $v = \frac{1}{6}$ ms^{-1}
24. a) 1 hr.

    b) 4 km.

    c) $v = 5.625$ kmh^{-1}
25. $s = t^3 - 2t^2 + 5t + 1$
26. $v = \frac{1}{4}t^4 - 3t^2 + 6$

    $s = \frac{1}{20}t^5 - t^3 + 6t + 5$
27. 18.25 m
28. $a = \frac{6}{\sqrt{4t+1}}$

    $s = \frac{(4t+1)^{\frac{3}{2}} + 7}{2}$
29. maximum velocity = 3 (in the negative direction);

    maximum displacement = 7 (in the negative direction)
30. i) A = 4; proof.

    ii) $s = 450 - \frac{3375}{t^t}m$

    iii) 5.4 ms^{-1}
31. i) $s = 25$ m

    ii) Proof.
32. i) $t = 8$ s

    ii) $s = 108$ m
33. i) $v = 0.3t^2$

    $s = 100$ m

    ii) 194 m

Exam-style paper A page 136

1. i) a) 0.495 N
   b) 14.7 N
   ii) 14.78 N, 88.1°
2. $\theta = 41.8°$
   X = 2.24 N
3. i) R = 7.89 N
   ii) $\mu = 0.291$
4. i) 0.2 ms^{-2}
   ii) 100 s
   iii) 5040 m
   iv) 4900 m
5. $h = 1$m
6. i) $t = 10.5$ s
   ii) $s = 408$ m
7. i) 20 ms^{-2}
   ii) 1040 m
8. i) 2.1 s ii) 11.55 m
9. i) 745 J ii) 149 N

Exam-style paper B page 138

1. 8.31 N, 26.3° below the positive x-axis
2. $P = 4.92$ N, $\theta = 74.6°$
3. i) $R = 3.23$ N
   ii) $F = 0.548$ N
   iii) 0.169
4. i)

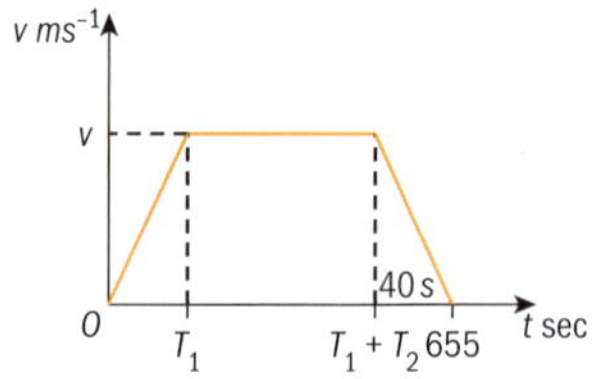

   ii) $T_1 = 2.5V$, deceleration = $\frac{V}{40}$
   iii) $V = 8$ ms^{-1}
5. i) $T = 5.4$ N
   ii) v = 2 ms^{-1}
   iii) 0.8 m
6. i) $a = 3$ ms^{-2}
   ii) $s = 5$ m
7. i) 1.28 ms^{-1} ii) 651 m
8. i) Proof ii) 22 ms^{-1}
9. i) 2820 J ii) 654 J iii) 2.29 ms^{-1}

Index

Page numbers in *italics* refer to questions sections.

A

acceleration 128
- acceleration and variable resistance 110–15
- acceleration due to gravity 50
- constant acceleration formulae 18, 19, 19–23, *28–9*, 120, 126
- uniform acceleration 18

aerodynamics 134–5
analogue machines 118

B

base jumping 97
Bernoulli Principle 135
bodies 4, 50
- equilibrium 63

bullet train 90

C

celestial mathematics 88–9
coefficient of friction 73
components 35–6, 40
conservation of energy 96–9
conservative systems 96
constant acceleration formulae 18, 19, 19–23, *28–9*, 120, 126
- answers to questions 141–42

Copernicus, Nicolaus 49, 88
cosines 32, 40
cricket 46

D

derivative 120
Descartes, René 49
differentiation 119–22, 128
displacement 19, 29, 128
displacement–time graphs 4–7, 17
distance 19, 29

E

Einstein, Albert 89
energy 88, *103–4*, 105
- answers to questions 146–7
- conservation of energy 96–9
- gravitational potential energy (GPE) 95–6
- kinetic energy 94–5
- work-energy principle 99–102

Englert, François 89
equilibrium 62, *67–9*
- answers to questions 144–5
- limiting equilibrium 73
- three forces acting at a point 63–6, 69

exam-style paper A 136–7
- answers to questions 149

exam-style paper B 138–9
- answers to questions 149

F

forces 30, *36–40*
- answers to questions 142
- components 35–6
- multiple forces 55–7
- resolved parts 35
- three forces acting at a point 63–6, 69
- triangle of forces 63

Formula One racing 106
free fall 24
friction 72–3, *81–82*, 83
- answers to questions 145–6
- coefficient of friction 73
- frictional forces 73
- rough horizontal surfaces 73–6
- rough inclined slope 76–80

funicular railways 59

G

Gagarin, Yuri 48
Galileo Galilei 50
graphs 2–3, *16–17*
- answers to questions 140–41
- displacement–time graphs 4–7, 17
- velocity-time graphs 8–15, 17

gravitational potential energy (GPE) 95–6
gravity 50, 89

H

Higgs, Peter 89

I

integration 123–6, 128

J

joule (J) 91, 96
Joule, James Prescott 91

K

Kepler, Johannes 49, 98
kilojoules 96
kinematics 4
kinetic energy 94–5

L

light inextensible strings 57
limiting equilibrium 73

M

modelling conditions 29

N

Newton, Isaac 24, 50, 88, 89
Newton's laws 48, 51–3, *60–1*
- answers to questions 144
- connected particles 57–9
- multiple forces 55–7
- Newton's first law 49, 61
- Newton's second law 49–50, 61
- Newton's third law 50, 61
- resolving when on an inclined plane 53–4, 61

non-conservative systems 96, 99
normal reaction 65, 73

P

parallelogram of forces 32
particles 4, 50
- connected particles 57–9
- Higgs-Boson 89

power 106–7, *115–17*, 117
- acceleration and variable resistance 110–15
- answers to questions 147
- power as rate of doing work 107–110, 117

pulleys 57

R

resolving 35
- resolving parallel to the surface of the slope 65, 77, 78

resolving perpendicular to the surface of the slope 65, 77–8, 79
resolving vertically 65, 75
resolving when on an inclined plane 53–4, 61
resultants 30, 31–5, *36–40*
answers to questions 142
resultant vector 31
review exercises 41–5, 84–7, 129–33
answers to questions 143–4, 146, 148
rough surfaces 72
rough horizontal surfaces 73–6
rough inclined slope 76–80

S

S.I. units 91, 117
scalar quantities 10–11, 19
sine rules 63
slopes 53, 65, 76–80
smooth pulleys 57
smooth surfaces 72
soccer 46
speed 19, 29
sports technology 46–7
straight line motion 2–3, *16–17*, 128
answers to questions 140–41
using differentiation to describe straight line motion 119–22
using integration to describe straight line motion 123–6
strings 57
surfaces 72, 73–6

T

tennis 46–7
Trans-Australian Railway 2
triangle of forces 63

U

uniform acceleration 18

V

variable forces 118, *127*, 128
answers to questions 147–8
constant acceleration formulae 126
using differentiation to describe straight line motion 119–22
using integration to describe straight line motion 123–6
variable resistance 110–15
vector quantities 10–11, 19
velocity 19, 29, 128
velocity-time graphs 8–15, 17
vertical motion 24–7, 29
sign convention 24–7

W

Watt, James 107
weightlessness 51
work 90, 91–93, *103–4*, 105
answers to questions 146–7
work done 91
work-energy principle 99–102